In Pursuit of Zeta-3

THE HERO OF THIS BOOK

Leonhard Euler (1707–1783), "analysis incarnate." 1753 portrait by Jakob
Emanuel Handmann. Image courtesy of the AIP Emilio Segrè Visual Archives.

In Pursuit of Zeta-3

The World's Most Mysterious
Unsolved Math Problem

PAUL J. NAHIN

PRINCETON UNIVERSITY PRESS
Princeton & Oxford

Published by Princeton University Press
41 William Street, Princeton, New Jersey 08540
99 Banbury Road, Oxford OX2 6JX

press.princeton.edu

First paperback printing, 2023
Paperback ISBN 9780691247649

The Library of Congress has cataloged the cloth edition as follows:

Names: Nahin, Paul J., author.
Title: In pursuit of zeta-3 : the world's most mysterious unsolved math
    problem / Paul J. Nahin.
Description: Princeton : Princeton University Press, [2021] | Includes bibliographical
    references and index.
Identifiers: LCCN 2021018589 (print) | LCCN 2021018590 (ebook) |
    ISBN 9780691206073 (hardback) | ISBN 9780691227597 (ebook)
Subjects: LCSH: Functions, Zeta. | Mathematics--Philosophy. | BISAC: MATHEMATICS
    / History & Philosophy | TECHNOLOGY & ENGINEERING / General
Classification: LCC QA351 .N34 2021 (print) | LCC QA351 (ebook) | DDC 515/.56—dc23
LC record available at https://lccn.loc.gov/2021018589
LC ebook record available at https://lccn.loc.gov/2021018590

British Library Cataloging-in-Publication Data is available

Editorial: Susannah Shoemaker and Kristen Hop
Production Editorial: Debbie Tegarden and Mark Bellis
Text and Jacket/Cover Design: Lauren Smith
Production: Jacqueline Poirier
Publicity: Matthew Taylor and Carmen Jimenez
Copyeditor: Cyd Westmoreland

This book has been composed in Adobe Text Pro and Arboria

Contents

3. Periodic Functions, Fourier Series, and the Zeta Function

4. Euler Sums, the Harmonic Series, and the Zeta Function

Other sciences seek to discover the laws that God has chosen; mathematicians seek to discover the laws that God has to obey.

—Raoul Bott (1923–2005), in an address to the Knesset as he accepted the Wolf Prize in Mathematics (2000)

Preface

This book is not a research monograph. Instead, it is an historical introduction to the nature of zeta-3 and related topics. By "historical," I mean I will mention the names of the mathematicians associated with the mathematics we'll discuss—I'll even mention some biographical details here and there—and will do my very best to get right any dates cited.

The book has been written with the hope that, in particular, enthusiastic readers of mathematics at the level of high school AP-calculus (that is, budding mathematicians, mathematical physicists, and research engineers) will form the primary audience. Such readers will be able to understand nearly everything in this book (with some exceptions that I'll mention next) and should then be well prepared for more advanced study of the zeta-3 problem.

What are those exceptions I just mentioned? This is an important issue for me to be explicit about, and for you to understand. Most of the mathematics you'll find in this book will be accessible to an AP-calculus student, but not all of it will be without some further study on your part. For example, there are places in the book where double integrals occur, and in such places, I have simply assumed that you'll get hold of a good calculus text and read up on multiple integrals if you need to do so. (When you've gotten into the book as far as Chapter 3, you should then read Appendix 4.) If you are interested

enough in mathematics to be reading this Preface, I think you are the
sort of person for whom doing that is not asking too much.

More broadly, the general approach of the book (written by an electrical engineer, not a mathematician) is to keep the exposition intuitive
and plausible, and so I have not provided detailed justifications for
some of the mathematical deductions made, justifications that a mathematician might desire. Be assured, however, that such deductions can
be justified rigorously, using established (if advanced) mathematics.

To enhance the usefulness of the book for self-study, numerous
challenge questions, at least one for nearly every section of the book,
with complete solutions at the end, are included. Some are easy,
others not so easy (but hints are provided to help get you started).
Now, a natural question to ask at this point is: Am I (that is, are **you**)
at a mathematical level to be able to read this book? In response to
that, here are some tests of increasing difficulty that will answer that
question.

First, there are a lot of summations in this book, and their evalua-
tions, while using only elementary mathematics, may require one or
more insightful observations. For example, suppose we define
$h(q) = \sum_{k=1}^{q} \frac{1}{k} = 1 + \frac{1}{2} + \frac{1}{3} + \cdots + \frac{1}{q}$, where q is any positive integer. As
you'll learn in Chapter 1, $\lim_{q \to \infty} h(q) = \infty$. Now, consider the sum
$S = \sum_{q=1}^{\infty} \frac{h(q)}{q}$. Do you see that $S = \infty$? Simply notice that $h(q) \geq 1$ for all
$q \geq 1$. Thus, if we replace $h(q)$ with 1 in the definition of S, we'll obvi-
ously get a smaller sum and so $S > \sum_{q=1}^{\infty} \frac{1}{q}$. But that sum blows up (you
know, since I just told you this a few sentences ago). Getting
$S > \infty$ is, of course, just an enthusiastic way of concluding $S = \infty$. Well,
that wasn't so hard, was it?

Okay, here's the second test. What is the value of $\int_0^1 \sqrt{x - x^2}\, dx$? If
you can actually do this definite integral, well, that's terrific. But that's
not the test. Even if you can't do it, as before, if you can understand the
following solution, then you are all set. The integral is the area under
the curve (and above the x-axis) described by $y(x) = \sqrt{x - x^2}$, $0 \leq x \leq 1$.
That is, by the curve $y^2 = x - x^2$, or $x^2 - x + y^2 = 0$ or, completing the
square, by the curve $(x - \frac{1}{2})^2 + y^2 = (\frac{1}{2})^2$. This is the equation of a circle
with radius $\frac{1}{2}$, centered on the x-axis at $x = \frac{1}{2}$. The area bounded by that

curve and the x-axis is, therefore, the area of the upper half of the circle. That is, $\int_0^1 \sqrt{x - x^2}\, dx = \frac{1}{2}\pi(\frac{1}{2})^2 = \frac{\pi}{8}$.

One troublesome thought you may be bothered by at this point is that these two tests could seem a bit contrived or "cooked up," perhaps even to be a bit too mathematically academic. If you're an engineer or a physicist, seeing mathematics in the service of physics might be more reassuring to you that the level of this book is relevant to real-world science. So, for a third test, to show you that with an AP-calculus background, you can understand some pretty sophisticated mathematical analysis—certainly anything you'll read in this book—consider the integral $\int_0^T \frac{e^{-\frac{a}{T-\tau} - \frac{b}{\tau}}}{(T-\tau)^{1/2}\,\tau^{3/2}}\, d\tau$, where T, a, and b are all positive constants (τ, in contrast, obviously varies from 0 to T). This certainly looks complicated (that's code for *awfully hard*), but in fact it yields (and quickly, too) in the face of nothing but AP-calculus.

Why, you ask, is this an interesting calculation? It's interesting because this integral appears at the end of a 1965 book by Richard Feynman (1918–1988) and Albert Hibbs (1924–2003), *Quantum Mechanics and Path Integrals* (Dover, 2010). That book discusses mathematical physics at the heart of the contributions to quantum electrodynamics that won Feynman a share of the 1965 Nobel Prize in physics.[1] Doing the Feynman-Hibbs integral isn't going to make you a quantum theory expert, but to become a quantum theory expert, you'll have to be able to understand how to do the integral. And that is something an AP-calculus student can do. We'll have to work out this integral for ourselves, as Feynman and Hibbs gave only the answer and no derivation. If you can track along through the following solution, then you are definitely all set to go forward with this book.

To start, for notational convenience (you'll see why as we go along), let's write τ, a, and b in terms of the constant T. That is, let's define the so-called *scaling parameters* x, α, and β to be such that

1. Feynman was a professor of physics at the California Institute of Technology, and Hibbs was a former graduate student at Caltech who had done his PhD dissertation under Feynman. I thank Anthony Zee, professor of physics at the University of California at Santa Barbara, who brought the Feynman-Hibbs integral to my attention.

$\tau = xT$, $a = \alpha T$, and $b = \beta T$. (Notice that α and β are constants, but since τ is a variable, then so, too, is x, with $d\tau = T\,dx$.) The Feynman-Hibbs integral then becomes cleaner looking:

$$\int_0^1 \frac{e^{-\frac{\alpha T}{T-xT}-\frac{\beta T}{xT}}}{(T-xT)^{1/2}(xT)^{3/2}}\,T\,dx = T\int_0^1 \frac{e^{-\frac{\alpha}{1-x}-\frac{\beta}{x}}}{T^2(1-x)^{1/2}\,x^{3/2}}\,dx$$

$$= \frac{1}{T}\int_0^1 \frac{e^{-\frac{\alpha}{1-x}-\frac{\beta}{x}}}{(1-x)^{1/2}\,x^{3/2}}\,dx.$$

(Note that since our integration variable is now x, not τ, the integration limits must be for x and so we changed the upper integration limit to 1 because $x = 1$ when $\tau = T$.) Next, change variable to $y = \frac{1}{x}$ (and so $dx = -\frac{dy}{y^2}$) to write the Feynman-Hibbs integral as

$$\frac{1}{T}\int_\infty^1 \frac{e^{-\frac{\alpha}{1-\frac{1}{y}}-\frac{\beta}{\frac{1}{y}}}}{\left(1-\frac{1}{y}\right)^{1/2}\left(\frac{1}{y}\right)^{3/2}}\left(-\frac{dy}{y^2}\right) = \frac{1}{T}\int_1^\infty \frac{e^{\frac{\alpha y}{y-1}-\beta y}}{y^2\frac{(y-1)^{1/2}}{y^2}}\,dy = \frac{1}{T}\int_1^\infty \frac{e^{\frac{\alpha y}{y-1}-\beta y}}{(y-1)^{1/2}}\,dy.$$

Now change variable to $u = y - 1$, and so the Feynman-Hibbs integral becomes

$$\frac{1}{T}\int_0^\infty \frac{e^{-\alpha\frac{u+1}{u}-\beta(u+1)}}{u^{1/2}}\,du = \frac{1}{T}\int_0^\infty \frac{e^{-\alpha\left(1+\frac{1}{u}\right)-\beta u-\beta}}{u^{1/2}}\,du$$

$$= \frac{1}{T}\int_0^\infty \frac{e^{-\alpha-\frac{\alpha}{u}-\beta u-\beta}}{u^{1/2}}\,du = \frac{1}{T}\int_0^\infty \frac{e^{-(\alpha+\beta)}e^{-\frac{\alpha}{u}-\beta u}}{u^{1/2}}\,du = \frac{e^{-(\alpha+\beta)}}{T}\int_0^\infty \frac{e^{-\frac{\alpha}{u}-\beta u}}{u^{1/2}}\,du.$$

Next, change variable one last time to $u = \omega^2$ (and so $du = 2\omega\,d\omega$). The Feynman-Hibbs integral then becomes

$$\frac{e^{-(\alpha+\beta)}}{T}\int_0^\infty \frac{e^{-\frac{\alpha}{\omega^2}-\beta\omega^2}}{\omega}\,2\omega\,d\omega = \frac{2e^{-(\alpha+\beta)}}{T}\int_0^\infty e^{-\frac{\alpha}{\omega^2}-\beta\omega^2}\,d\omega.$$

This last integral may look like all we've done is trade the Feynman-Hibbs integral for a new but equally awful one, but that's not so. In fact, in Appendix 2, I work out this last integral (using only AP-calculus arguments) as an extension of the famous "probability integral" $\int_0^\infty e^{-t^2} dt = \frac{1}{2}\sqrt{\pi}$, first done 250 years ago (also using only AP-calculus level arguments), which we'll use in Chapter 1 when we get to the gamma function. Specifically, in Appendix 2 you'll find the derivation of the formula

$$\int_0^\infty e^{-pt^2 - \frac{q}{t^2}} dt = \frac{1}{2}\sqrt{\frac{\pi}{p}} e^{-2\sqrt{pq}}$$

where p and q are non-negative but otherwise arbitrary real constants (this reduces to the probability integral for the case of $p = 1$ and $q = 0$).

For our calculation, $p = \beta = \frac{b}{T}$ and $q = \alpha = \frac{a}{T}$, and so the Feynman-Hibbs integral is

$$\int_0^T \frac{e^{-\frac{a}{T-\tau}-\frac{b}{\tau}}}{(T-\tau)^{1/2}\,\tau^{3/2}} d\tau = \left\{\frac{2e^{-\left(\frac{a}{T}+\frac{b}{T}\right)}}{T}\right\}\left\{\frac{1}{2}\sqrt{\frac{\pi}{\frac{b}{T}}}e^{-2\sqrt{\frac{ab}{T^2}}}\right\} = \sqrt{\frac{\pi}{bT}}e^{-\left(\frac{a}{T}+\frac{b}{T}\right)-2\frac{\sqrt{ab}}{T}}$$

$$= \sqrt{\frac{\pi}{bT}}e^{-\frac{1}{T}(a+2\sqrt{ab}+b)} = \sqrt{\frac{\pi}{bT}}e^{-\frac{1}{T}(\sqrt{a}+\sqrt{b})^2},$$

which is the answer given by Feynman and Hibbs.

Well, that was a lot of pretty math—*but is it right?* We can greatly enhance our confidence in this theoretical result by numerically evaluating the original integral for specific values of T, a, and b and comparing the result with what our formula says (we'll do this sort of checking a lot throughout this book). For example, if $a = b = 1$ and $T = 4$, the theoretical answer reduces to $\frac{\sqrt{\pi}}{2e} = 0.3260246660866$. Using modern integration software commercially available for personal computers (in this book, I use *MATLAB*), typing the single line[2]

$$integral(@(x)(exp(-1./(4-x)).*exp(-1./x))./$$
$$sqrt((x.^3).*(4-x)),0,4)$$

almost instantly returns the value $0.3260246660868\ldots$, in excellent agreement with theory. (In this line, x plays the role of the original integration variable τ.)

For your penultimate test, let's return to the $h(q)$ function introduced earlier. That summation function appears, in a remarkable way, in the evaluation of the integral $\int_0^1 x^{2q-1}\ln(1+x)dx$, where q is any positive integer. We start by making the observation that

$$x^{2q-1} = \frac{d}{dx}\left(\frac{x^{2q}}{2q} - \frac{1}{2q}\right),$$

and so

$$\int_0^1 x^{2q-1}\ln(1+x)dx = \int_0^1 \left\{\frac{d}{dx}\left(\frac{x^{2q}}{2q} - \frac{1}{2q}\right)\right\}\ln(1+x)dx$$

$$= \int_0^1 \ln(1+x)d\left(\frac{x^{2q}}{2q} - \frac{1}{2q}\right).$$

This observation is, I think, not an obvious one to make, although once seen, it is obviously true.[3] The reason for doing this becomes clear once you recall the AP-calculus integration-by-parts formula

---

2. The format of the *MATLAB* command *integral* is straightforward: integrand, lower limit, upper limit. I show you this line of code not because I'm trying to turn you into a *MATLAB* coder, but simply as an example of how easy it is to use a computer as an effective tool in what a relatively short time ago would have been considered to be a purely theoretical math problem.

3. I learned this little trick from a note sent to me by Cornel Ioan Vălean, the author of *(Almost) Impossible Integrals, Sums, and Series* (Springer, 2019).

$$\int_0^1 u \, dv = (uv)\big|_0^1 - \int_0^1 v \, du$$

and make the associations $u = \ln(1 + x)$ (and so $du = \frac{1}{1+x}dx$), and $dv = d(\frac{x^{2q}}{2q} - \frac{1}{2q})$ (and so $v = \frac{x^{2q}}{2q} - \frac{1}{2q}$). Then

$$\int_0^1 x^{2q-1}\ln(1+x)dx = \ln(1+x)\left(\frac{x^{2q}}{2q} - \frac{1}{2q}\right)\Big|_0^1 - \int_0^1\left(\frac{x^{2q}}{2q} - \frac{1}{2q}\right)\frac{1}{1+x}dx$$

$$= \frac{1}{2q}\int_0^1 \frac{1-x^{2q}}{1+x}dx.$$

Now, as

$$\frac{1}{1+x} = 1 - x + x^2 - x^3 + x^4 - \cdots$$

then

$$\frac{1-x^{2q}}{1+x} = (1-x^{2q})(1 - x + x^2 - x^3 + x^4 - \cdots)$$
$$= (1 - x + x^2 - x^3 + \cdots) - (x^{2q} - x^{2q+1} + x^{2q+2} - x^{2q+3} + \cdots).$$

That is,

$$\frac{1-x^{2q}}{1+x} = [1 - (sum \ of \ all \ odd \ powers \ of \ x) + (sum \ of \ all \ even \ powers \ of \ x)]$$

$$-\left[\begin{array}{l}(sum \ of \ even \ powers \ of \ x \ starting \ with \ x^{2q}) \\ -(sum \ of \ odd \ powers \ of \ x \ starting \ with \ x^{2q+1})\end{array}\right]$$

$$= 1 + (sum \ of \ even \ powers \ of \ x \ from \ x^2 \ to \ x^{2q-2})$$

$$- (sum \ of \ odd \ powers \ of \ x \ from \ x^1 \ to \ x^{2q-1})$$

$$= 1 - x + x^2 - x^3 + x^4 - \cdots + x^{2q-2} - x^{2q-1} = \sum_{k=1}^{2q} (-1)^{k-1} x^{k-1}.$$

So

$$\int_0^1 x^{2q-1} \ln(1+x)\,dx = \frac{1}{2q} \int_0^1 \sum_{k=1}^{2q} (-1)^{k-1} x^{k-1}\,dx$$

or, reversing the order of integration and summation (a step that is immediately justified, since the sum has a finite number of terms), we have

$$\int_0^1 x^{2q-1} \ln(1+x)\,dx = \frac{1}{2q} \sum_{k=1}^{2q} (-1)^{k-1} \int_0^1 x^{k-1}\,dx = \frac{1}{2q} \sum_{k=1}^{2q} (-1)^{k-1} \left(\frac{x^k}{k} \right)\Big|_0^1$$

$$= \frac{1}{2q} \sum_{k=1}^{2q} (-1)^{k-1} \frac{1}{k} = \frac{1}{2q} \left[1 - \frac{1}{2} + \frac{1}{3} - \frac{1}{4} + \frac{1}{5} - \cdots + \frac{1}{2q-1} - \frac{1}{2q} \right]$$

$$= \frac{1}{2q} \left[\left(1 + \frac{1}{3} + \frac{1}{5} + \cdots + \frac{1}{2q-1} \right) - \left(\frac{1}{2} + \frac{1}{4} + \cdots + \frac{1}{2q} \right) \right]$$

$$= \frac{1}{2q} \left[\left(1 + \frac{1}{3} + \cdots + \frac{1}{2q-1} \right) + \left(\frac{1}{2} + \frac{1}{4} + \cdots + \frac{1}{2q} \right) - \left(\frac{1}{2} + \frac{1}{4} + \cdots + \frac{1}{2q} \right) - \left(\frac{1}{2} + \frac{1}{4} + \cdots + \frac{1}{2q} \right) \right]$$

$$= \frac{1}{2q} \left[\left(1 + \frac{1}{2} + \frac{1}{3} + \frac{1}{4} + \cdots + \frac{1}{2q-1} + \frac{1}{2q} \right) - 2 \left(\frac{1}{2} + \frac{1}{4} + \cdots + \frac{1}{2q} \right) \right]$$

$$= \frac{1}{2q} \left[h(2q) - \left(1 + \frac{1}{2} + \cdots + \frac{1}{q} \right) \right]$$

and so, at last, we arrive at the very pretty

$$\int_0^1 x^{2q-1} \ln(1+x)\,dx = \frac{h(2q) - h(q)}{2q}, \quad q \geq 1.$$

We'll run into more expressions (called *Euler sums*) that have forms similar to the right-hand side of this result when we get to Chapter 4. There you'll see that many of the Euler sums have values that are intimately related to zeta-3 (as the title proclaims, zeta-3 is the whole point of this book—but that's getting a bit ahead of ourselves). This is a pretty sophisticated analysis, but there is not a single step in it that isn't part of AP-calculus, and if you can follow it, you pass this test. Just to convince you that the analysis we did is okay, suppose $q = 3$. Then our result says

$$\int_0^1 x^5 \ln(1+x)dx = \frac{h(6)-h(3)}{6} = \frac{\left(1+\frac{1}{2}+\frac{1}{3}+\frac{1}{4}+\frac{1}{5}+\frac{1}{6}\right)-\left(1+\frac{1}{2}+\frac{1}{3}\right)}{6} = \frac{1}{6}\left(\frac{1}{4}+\frac{1}{5}+\frac{1}{6}\right)$$

$$= \frac{1}{6}\left(\frac{30+24+20}{120}\right)$$

$$= \frac{74}{720} = 0.102777\ldots,$$

while *MATLAB* says *integral(@(x)(x.^5).*log(1+x),0,1) = 0.102777* . . . , which is pretty good agreement.

As a final comment on this calculation, notice that we have, as a by-product of doing an integral, derived the identity

$$1 - \frac{1}{2} + \frac{1}{3} - \frac{1}{4} + \cdots - \frac{1}{2q} = h(2q) - h(q).$$

This result is famous in mathematics as the *Botez-Catalan identity,*[4] and we derived it by direct algebraic manipulation. In the final challenge question of this Preface, you'll be asked to do it in a different way.

Finally, here's a test of what some analysts like to call "mathematical maturity" (that's code for "able to follow a logical argument," not just one of symbol pushing). Consider the infinite sequence of positive integers defined by the recursion $x_k = x_{k-1} + x_{k-2}$, where $x_1 = x_2 = 1$. That is, consider the sequence 1,1,2,3,5,8,13,21, . . . , where each new integer is the sum of the previous two. (You might recognize this as the famous *Fibonacci sequence,* named after the 13th-century Italian mathematician Leonardo of Pisa, who wrote under the name of Fibonacci, a contraction of "filius Bonacci," that is, "the son of [Guglielmo] Bonacci.") The numbers in this sequence get pretty big,

---

4. After the Romanian mathematician and civil engineer Stefan Botez (1843–1920) and the French mathematician Eugène Catalan (1814–1894).

pretty fast (x_{50} = 12,586,269,025 ≈ 1.25 × 10^{10}), but there is a simple upper bound on each of them: $x_k < 2^k$. Here's how to show that, using the powerful analysis tool of *induction*. (Note that 2^{50} ≈ 1.12 × 10^{15}, and so our bound is pretty loose!)

First, we observe that the claim is certainly true for $k = 1$ and $k = 2$, as $x_1 = 1 < 2^1 = 2$ and $x_2 = 1 < 2^2 = 4$. We next show that if we *assume* the claim is true for $k = n$ and $k = n - 1$, it then follows that the claim is true for $k = n + 1$. If we can do that, then since the claim is true for $k = 1$ and $k = 2$, it also must be true for $k = 3$. Since we then know it's true for $k = 2$ and $k = 3$, it also must be true for $k = 4$. And so on, forever. Here are the details of showing this.

Our assumption that the claim is true for $k = n$ and $k = n - 1$ means that $x_n < 2^n$ and $x_{n-1} < 2^{n-1}$. Now, as $x_{n+1} = x_n + x_{n-1}$, it follows that $x_{n+1} < 2^n + 2^{n-1} = 2^n + 2^{-1}2^n = 2^n(1+\frac{1}{2}) < 2^n(2) = 2^{n+1}$, and we are done. This analysis involves no advanced, sophisticated math, but it does require the "maturity" I mentioned earlier.

If all these tests make sense, then keep reading. But first, try your hand at your first three challenge questions: First, show that $\int_{-\infty}^{\infty} e^{-ax^2+bx}dx = \sqrt{\frac{\pi}{a}}e^{b^2/4a}$, $a > 0$. Hint: Try completing the square, and recall the probability integral. Second, explain why $\int_0^1 \frac{1}{\sqrt{x-x^2}}dx = \pi$. In fact, this is the $a = 0$, $b = 1$, special case of $\int_a^b \frac{1}{\sqrt{(x-a)(b-x)}}dx = \pi$ for any non-negative values of $b > a$. See if you can explain *that*. And finally, see if you can establish the Botez-Catalan identity by induction. Hint: You'll find it helpful to notice that $h(q+1) = h(q) + \frac{1}{q+1}$.

Euler's Problem

1.1 Introducing Euler

The title of this book has been carefully crafted to attract the interest of all those who love mathematics, which would seem to be an obvious thing to do for the author of a book like this one. However, the subtitle of this book seems likely to provoke controversy among professional mathematicians, which, at the other extreme, might seem to be a rather odd thing for an author to do. My primary goal is clear, I think, as everybody likes a good hunt involving puzzles, a fact that explains the attraction of mystery novels, adventure video games, and *Indiana Jones* movies like *Raiders of the Lost Ark*. Hardly anybody, I think, would quibble with that. But how, I can hear each mathematician on the planet grumbling as he/she reads this, can I claim that the puzzle of zeta-3—I'll tell you what *that* is, in just a bit—is the world's most puzzling unsolved math problem? After all, as each of my critics would emphatically state, even while (perhaps) vigorously pounding a desktop or thumping a finger into my chest, "it's quite clear that the problem that's holding *my work* up is the world's most puzzling unsolved problem!"

My selection criteria for choosing which math problem is assigned the label as the most puzzling problem are quite simple: (1) the problem is (obviously!) unsolved; (2) people have been trying (and failing) for centuries to solve it; (3) it has at least some connection to the real world of physics and engineering; and most important of all, (4) despite (1), (2), and (3), a grammar school student who knows how to do elementary arithmetic can instantly understand the problem. The first three criteria are satisfied by lots of really hard problems in math, but if it takes a degree in math to simply understand the question, then such problems clearly fail the fourth test (this eliminates the famous problem of the Riemann hypothesis, about which I'll say more in the next section). At the end of the next chapter, I'll return to this issue, that of selecting the most puzzling math problem.[1]

But for now, let me set the stage for all that follows by introducing the personality most closely associated with the problem of zeta-3, the great (perhaps the *greatest* in history) mathematician, Leonhard Euler (1707–1783). The son of a rural Swiss pastor, Euler trained for the ministry at the University of Basel and at age 17, received a graduate degree from the Faculty of Theology. While a student at Basel, however, he also studied with the famous mathematician Johann Bernoulli (1667–1748), and despite his years-long immersion in religious thought, it was mathematics that captured his soul. Euler never lost his belief in God and in an afterlife, but while he was in *this* world, it was mathematics that had his supreme devotion.

It seemed that there was nothing that could keep him from doing mathematics, not even blindness from a botched cataract operation.

1. Until it was solved in 1995 by Andrew Wiles, perhaps Fermat's Last Theorem would be the problem that would have occurred to most people as the "world's most puzzling math problem," even though many professional mathematicians would have disagreed: for example, the great German mathematician Carl Friedrich Gauss (1777–1855), perhaps as great as Euler, refused to work on the Fermat problem, because he simply found it uninteresting. And, unlike the zeta-3 problem, the Fermat problem makes no appearance (as far as I know) in either science or engineering. Finally, the 1995 solution has been examined and understood by only a few world-class mathematicians. Everybody else simply accepts their thumbs-up verdict that Wiles' proof is correct (it's certainly far beyond AP-calculus!).

(Can you imagine enduring, with no anesthetic, such an operation in the 18th century?) Euler had a marvelous memory (it was said he knew the thousands of lines in the *Aeneid* by heart) and so, for the last 17 years of his life after losing his vision, he simply did monstrously complicated calculations in his head and dictated the results to an aide. Many years after his death, the 19th-century French astronomer Dominique Arago said of him, "Euler calculated without apparent effort, as men breathe or as eagles sustain themselves in the wind." By the time he died, he had written more brilliant mathematics than had any other mathematician in history, and that claim remains true to this day.

Here's one of Euler's accomplishments. Some of the great problems of mathematics involve the prime numbers, which since Euclid's day (more than three centuries before Christ) have been known to be infinite in number. Euclid's proof of that is a gem, commonly taught in high school (see the box), and it wasn't until 1737 that Euler found a second, totally different (but equally beautiful) proof of the infinity of the primes that I'll show you later in this chapter. But what Euler wasn't able to prove (and nobody else since has either) is if the *twin primes* are infinite in number. With the lone exception of 2, all the primes are odd numbers, and two primes form a twin pair if they are consecutive odd numbers (3 and 5, or 17 and 19, for example). Mathematicians would be absolutely astounded if the twin primes are not infinite in number, but there is still no proof of that.[2]

---

2. This just goes to show that there will never be an end to wonderful math problems, because, if in the (most unlikely) event that the twin primes are someday shown to be finite in number, the hunt would then immediately begin for the largest pair! In 1919, the Norwegian mathematician Viggo Brun (1885–1978) showed that the sum of the reciprocals of the twin primes is finite:

$(\frac{1}{3} + \frac{1}{5}) + (\frac{1}{5} + \frac{1}{7}) + (\frac{1}{11} + \frac{1}{13}) + (\frac{1}{17} + \frac{1}{19}) + \cdots \approx 1.90216\ldots$, a number called *Brun's constant*. This small value does not, however, prove that there are a finite number of twin primes, but only that they thin out pretty fast. In 2013 the Chinese-born American mathematician Yitang Zhang showed (when at the University of New Hampshire, just down the hall from my old office in Kingsbury Hall) that there is an infinity of pairs of primes such that each pair is separated by no more than 70 million. In 2014 that rather large gap was reduced to 246. If it could be reduced to 2 (or shown it *couldn't* be so reduced), then the twin prime problem would be resolved.

Euclid proved the primes are infinite in number by showing that a listing of any finite number n of primes must necessarily be incomplete, and so there must instead be an infinite number of primes. Here's how the logic goes. Let the n listed primes be labeled $p_1, p_2, p_3, \ldots, p_n$. Then, consider the number $N = p_1 p_2 p_3 \cdots p_n + 1$, which is obviously not equal to any of the primes on the list. Now, N is either prime or it isn't. If it is then we have directly found a prime not on the list. If, however, N is not a prime, that means it can be factored into a product of two (or more) primes. Equally obvious, however, is that p_1 doesn't divide N (because of that +1), and in fact none of the rest of the primes on the list divides N either, for the same reason. The immediate conclusion is that there must be at least *two* more primes that are not on the list. Since this argument holds for any listing of finite length, there must, in fact, be an infinite number of primes. Done! You're not going to find many proofs in math more elegantly concise than that.

The problem of determining the size of the set of the twin primes is an unsolved problem that definitely fits most (if not all, as perhaps some extra explanation would be required for a grammar school student[3]) of my selection criteria. So, why (you ask) doesn't the twin prime problem deserve to have the label of being the most puzzling math problem? Well, maybe it does, but I'm making a judgment call here, with the following reason for why I've come down on the side of zeta-3. The twin prime problem appears to stand mostly alone, with few peripheral connections to the rest of math and science. In contrast, the zeta-3 problem is at the center of all sorts of other problems. (You'll see some of them, starting in the next section when I'll *finally* tell you what the zeta-3 problem is!) It's this issue, of the

3. For example, to understand the nature of the primes, it is necessary to first study the so-called *unique factorization theorem*, which says that every integer can be factored into a product of primes in exactly one way. This is not terribly difficult to show, but it is a step beyond mere arithmetic.

relative connectedness to the rest of math, that makes our ignorance of the nature of zeta-3 the more exasperating (hence, the more puzzling and mysterious) in comparison to the problem of the infinity (or not) of the twin primes.

Challenge Problem 1.1.1: In a 1741 letter to a friend, Euler made the following claim:

$\frac{2^{\sqrt{-1}}+2^{-\sqrt{-1}}}{2} \approx \frac{10}{13}$, a claim that must have appeared to his friend to be like something he would have found in a book of magical incantations. Calculate each side of this "almost equality" out to several decimal places and so verify Euler's claim. Hint: You may find what today is called *Euler's identity* to be of great help: $e^{ix} = \cos(x) + i\sin(x)$, where $i = \sqrt{-1}$. You can find an entire book on this identity in my *Dr. Euler's Fabulous Formula* (Princeton University Press, 2017), but you do not have to read that book to do this problem. Simply notice that $2^i = e^{\ln(2^i)} = e^{i\ln(2)}$ (and similarly for 2^{-i}). Then apply Euler's identity.

1.2 The Harmonic Series and the Riemann Zeta Function

As Euler entered the second half of his third decade, he was known to his local contemporaries as a talented mathematician, but to become a famous mathematician, it was necessary (as it is today) to be the first to solve a *really hard* problem. There are always numerous such problems in mathematics, but in the 1730s, there was one that was particularly challenging, one that satisfies all of my selection criteria. This was the problem of summing the infinite series of the reciprocals of the squares of the positive integers. That is, the calculation of

$$(1.2.1) \qquad \sum_{k=1}^{\infty} \frac{1}{k^2} = 1 + \frac{1}{2^2} + \frac{1}{3^2} + \frac{1}{4^2} + \cdots = ?$$

It's important to understand what is really being asked for in (1.2.1). The numerical value of the sum is a calculation in arithmetic (but it's not a trivial one, if one wants, for example, the first 1,000 correct digits), and almost from the very day the problem was first posed, it was known that the value is about 1.6 or so. But that's not what mathematicians wanted. They wanted an exact *symbolic expression* involving integers (and roots of integers), simple functions (like the exponential, logarithmic, factorial, and trigonometric), and known constants like π and e. The simpler that expression, the better, and in fact, Euler found such an expression in 1734. A little later I'll show you his brilliant solution (and not to keep you in suspense, run $\pi^2/6$ through your calculator). For now, my central point is that, from 1734 on, Euler was a superstar in mathematics whose fame extended from one end of Europe to the other. The origin of the problem in (1.2.1) played a big, continuing role in both Euler's life and the zeta-3 problem (which I admit I still have yet to tell you about, but I will, soon!).

In the 14th century, a similar problem had bedeviled mathematicians: summing the infinite series of the reciprocals of the positive integers. That is, calculating

$$(1.2.2) \qquad \sum_{k=1}^{\infty} \frac{1}{k} = 1 + \frac{1}{2} + \frac{1}{3} + \frac{1}{4} + \cdots = ?$$

Then, about 1350, the French mathematician and philosopher Nicole Oresme (c. 1320–1382) showed that the answer is infinity! That is, as mathematicians put it, the sum in (1.2.2) diverges. Oresme's claim, without exception, surprises (greatly!) students when they first are told this, because the individual terms continually get smaller and smaller (indeed, they are approaching zero). It just seems impossible that, eventually, if you keep adding these ever-decreasing terms, the so-called *partial sum* will exceed any value you wish. That is, no matter how large a number N that you name, there is a finite value for q such that

$$(1.2.3) \qquad h(q) = \sum_{k=1}^{q} \frac{1}{k} > N.$$

The symbol h is used in (1.2.3) because the sum of the reciprocals of the positive integers is called the *harmonic series*. The $h(q)$ function will occur over and over in this book. Some of Euler's most beautiful discoveries after 1734 involve $h(q)$, and it continues to inspire researchers to this day.

Oresme's proof of (1.2.3) is an elegant example of the power of mathematical reasoning, even at the high school level. One simply makes clever use of brackets to group the terms as follows:

$$\sum_{k=1}^{\infty}\frac{1}{k}=1+\left\{\frac{1}{2}\right\}+\left\{\frac{1}{3}+\frac{1}{4}\right\}+\left\{\frac{1}{5}+\frac{1}{6}+\frac{1}{7}+\frac{1}{8}\right\}+\cdots$$

followed by replacing each term in each pair of curly brackets with the last (*smallest*) term in that pair. Notice that this last term is always of the form $\frac{1}{2^m}$ where m is some integer ($m = 1$ in the first pair, $m = 2$ in the second pair, $m = 3$ in third pair, and so on), and that there are 2^{m-1} terms in a bracket pair. The process gives a *lower* bound on the sum, and so we have

$$\sum_{k=1}^{\infty}\frac{1}{k}>1+\left\{\frac{1}{2}\right\}+\left\{\frac{1}{4}+\frac{1}{4}\right\}+\left\{\frac{1}{8}+\frac{1}{8}+\frac{1}{8}+\frac{1}{8}\right\}+\cdots=1+\frac{1}{2}+\frac{1}{2}+\frac{1}{2}+\cdots.$$

That is, the lower bound is 1 plus an infinity of $\frac{1}{2}$'s, which obviously gives a sum that "blows up" (diverges) to infinity, just as claimed in the Preface:

(1.2.4) $\lim_{q\to\infty}h(q)=\infty.$

The explanation for (1.2.3) and (1.2.4) is that while it is clearly necessary for the terms in an infinite series in which every term is positive to continually decrease toward zero if the sum is to be finite (for the sum to *converge*, as mathematicians put it), a decrease alone is not a sufficient condition for a finite sum. Not only must the terms decrease toward zero, but that decrease has to be a sufficiently fast one. The terms of the harmonic series simply don't go to zero fast enough. *Almost* fast enough, to be sure, but not quite fast enough, which results in the divergence of the harmonic series being astonishingly slow. For $h(q) > 15$, for example, we must have $q > 1.6 \times 10^6$ terms, while $h(q) > 100$ requires $q > 1.5 \times 10^{43}$ terms.

Once Oresme had solved the problem of summing the harmonic series, the question of summing the reciprocals *squared* stepped forward, with its explicit statement attributed to the Italian Pietro Mengoli (1625–1686) in 1644. And once Euler had solved that problem in 1734, you can surely understand the curiosity that drove mathematicians to next turn their attention to summing the reciprocals *cubed*. To their dismay, they couldn't do it. Even Euler couldn't do it. And so, at last, we have the zeta-3 problem: What is

$$(1.2.5) \qquad \sum_{k=1}^{\infty} \frac{1}{k^3} = 1 + \frac{1}{2^3} + \frac{1}{3^3} + \frac{1}{4^3} + \cdots = ?$$

The numerical value is easily calculated to be 1.2020569 ... but, unlike the sum of the reciprocals squared, there is no known simple symbolic expression. The search for such an expression is, today, an ongoing effort involving many of the best mathematicians in the world.

This search is not an idle one of mere curiosity, either, as the value of zeta-3 appears in physics (as you'll see later) as well as in mathematics.

The pressure on modern academics to solve problems is, as it was in Euler's day, enormous, and in fact, that pressure is relentless. That is, after solving a tough problem, the successful analyst certainly gets a pat on the back but then, almost immediately after, is asked "So, what are you going to do next?" Having a good answer to that question may be more a matter of professional pride for a tenured senior professor, but for a young untenured assistant professor, it is, quite literally, a matter of survival. The famous Hungarian mathematician Paul Erdös (1913–1996) wrote a little witticism that nicely sums up this situation:

> A theorem a day
> Means promotion and pay.
> A theorem a year
> And you're out on your ear!

To that, in the spirit of this book, I would add these two lines:

> But if your next theorem computes zeta-3
> Then acclaimed tenured full prof you'll instantly be!

Erdös, who received the 1983 Wolf Prize, never held an academic position, but instead endlessly traveled the world, living temporarily with mathematician friends, then moving on to his next stop. At each stay, he and his host would write a joint paper (his co-authors numbered in the hundreds): his motto was "Another roof, another proof."

The reason for the name *zeta* is that in 1737, Euler considered the general problem of summing the reciprocals of the sth power of the positive integers:

$$(1.2.6) \qquad \sum_{k=1}^{\infty} \frac{1}{k^s} = 1 + \frac{1}{2^s} + \frac{1}{3^s} + \frac{1}{4^s} + \cdots,$$

which is today written (with the Greek letter zeta) as $\zeta(s)$, and so (1.2.1) is $\zeta(2)$ and (1.2.5) is $\zeta(3)$. That is, zeta(2) and zeta(3), pronounced "zeta-2" and "zeta-3." Euler took s to be a positive integer, subject only to the constraint that $s > 1$ to ensure convergence of (1.2.6) ($s = 1$ gives, of course, the divergent harmonic series). What makes the failure to solve the zeta-3 problem particularly puzzling is that not only did Euler solve the zeta-2 problem, but he also solved all of the zeta-2n problems. That is, he found symbolic expressions for the sum of the reciprocals of *any even* power of the integers. The first few of these solutions are:

$$\text{Zeta-2} = \zeta(2) = \frac{\pi^2}{6},$$

$$\text{Zeta-4} = \zeta(4) = \frac{\pi^4}{90},$$

$$\text{Zeta-6} = \zeta(6) = \frac{\pi^6}{945},$$

$$\text{Zeta-8} = \zeta(8) = \frac{\pi^8}{9,450},$$

$$\text{Zeta-10} = \zeta(10) = \frac{\pi^{10}}{93,555}.$$

Starting with $\zeta(3)$, however, not even one of the $\zeta(2n + 1)$ problems has been solved. Lots of results that dance around $\zeta(2n + 1)$ have been found since Euler—in 1979, for example, the French mathematician Roger Apéry (1916–1994) showed that, whatever $\zeta(3)$ is, it is irrational (which confirmed what every mathematician since Euler has always believed, but having a proof is, of course, the Holy Grail of mathematics).[4] A simple symbolic expression for $\zeta(3)$ remains as elusive today as it was for Euler.

---

4. The irrationality of zeta-2 wasn't proven until 1796, decades after Euler calculated $\zeta(2)$, when the French mathematician Adrien-Marie Legendre (1752–1833) proved that π^2 is irrational (the Swiss mathematician Johann Lambert (1728–1777) proved that π is irrational in 1761, but that does not prove that π^2 is irrational). Can you think of an irrational number whose square is rational? This should take you, at most, two (big hint here) seconds!

In 1859 the German mathematician Bernhard Riemann (1826–1866) extended Euler's $\zeta(s)$ to complex values of s (which, of course, includes the integers as special cases). Today, $\zeta(s)$ is called the *Riemann zeta function*, although its origin is with Euler. For highly technical reasons, beyond the level of this book, there are many important problems in mathematics (including the theory of primes) that are connected to what are called the *zeros* of $\zeta(s)$. That is, to the solutions of the equation $\zeta(s) = 0$. All the even, negative integer values of s are zeros, but the situation for complex zeros is far from resolved. After calculating just the first three (!) complex zeros, Riemann conjectured, but was unable to prove (and nobody else since has, either), that all the infinite number of complex zeros of $\zeta(s)$ are "very likely" of the form $s = \frac{1}{2} + b\sqrt{-1}$ for an infinite number of values for $b > 0$. That is, what has become known as the *Riemann hypothesis* is that all of the complex zeros are on the vertical line (called the *critical line*) in the complex plane with its real part equal to $\frac{1}{2}$. (In Chapter 3, I'll tell you a lot more about the critical line.) In 1914 the English mathematician G. H. Hardy (1877–1947) proved that $\zeta(s)$ has an infinite number of complex zeros on the critical line, but that does not prove that *all* the complex zeros are there. In 1989 it was shown that at least two-fifths of the complex zeros are on the critical line. Again, that does not prove that all the complex zeros are there. In 2011, 22 years later, that 40% value was increased to 41.05%, a small increase for two decades of work that hints at just how difficult a challenge the Riemann hypothesis is. Using high-speed electronic computers, billions upon billions of the complex zeros have been calculated as the parameter b is steadily increased and, so far, every last one of them does indeed have a real part of exactly $\frac{1}{2}$. But that does not say anything about all of the complex zeros being on the critical line. If just *one* complex zero is ever found off the critical line, then the Riemann hypothesis will be instantly swept into the wastebasket of history (and the discoverer of that rogue zero will become an instant superstar in the world of mathematics).

Euler's results for $\zeta(2n)$ all have the form of

$$\zeta(2n) = \frac{a}{b}\pi^{2n},$$

where a and b are positive integers. That is, for k an even integer, $\zeta(k)$ is a rational number, times pi to the kth power. This suggests that, for some integers a and b,

$$\zeta(3) = \frac{a}{b}\pi^3$$

but that suggestion has not been realized (nobody has ever found integers a and b that give the known numerical value of $\zeta(3)$). In 1740 Euler conjectured that, instead,

$$\zeta(3) = N\pi^3$$

where N somehow involves $\ln(2)$, but that hasn't resulted in any progress, either.

Why $\ln(2)$? Why not $\ln(17)$ or $\ln(3)$? Perhaps because Euler had shown, before he solved the zeta-2 problem, that $\zeta(2) = \sum_{k=1}^{\infty} \frac{1}{k^2 2^{k-1}} + \{\ln(2)\}^2$. This was helpful in calculating the *numerical* value of $\zeta(2)$ because this sum converges much more rapidly than does the original sum in the definition of $\zeta(2)$, and he knew the value of $\ln(2)$ to many decimal places. The Russian mathematician Andrei Markov (1856–1922) did the same for $\zeta(3)$ when he showed, in 1890, that $\zeta(3) = \frac{5}{2}\sum_{k=1}^{\infty}(-1)^{k-1}\frac{1}{k^3\binom{2k}{k}} = \frac{5}{2}\sum_{k=1}^{\infty}(-1)^{k-1}\frac{(k!)^2}{k^3(2k)!}$, where $\binom{n}{k}$ is the binomial coefficient $\frac{n!}{(n-k)!k!}$. I'll show you the details of how Euler derived his fast-converging series expression for $\zeta(2)$ (it's all just AP-calculus) in the next chapter. Markov's analysis is, as you might suspect, *just a bit* more advanced.

Another formula, one known to Euler (and which we'll derive later), is particularly tantalizing:

$$1 - \frac{1}{3^3} + \frac{1}{5^3} - \frac{1}{7^3} + \cdots = \frac{\pi^3}{32}.$$

Why the sum of the reciprocals of the *odd* integers cubed, with alternating signs, has such a nice form, a form that matches that of $\zeta(2n)$, while $\zeta(3)$ seems not to, is an exasperating puzzle for mathematicians. What worries many of today's mathematicians is that if *Euler*—a genius of the first rank (if not even higher)—couldn't solve for zeta-3, even after decades of trying, well, maybe there simply isn't an exact symbolic expression. What a dreary thought! Why would the world be made that way? It seems so . . . inelegant. And yet, such things *do* happen. The ancient geometric construction problems of angle trisection, cube doubling, and circle squaring, for example, all stumped mathematicians for thousands of years until all were eventually proven to have (using only a straightedge and a compass) no solutions (see Appendix 1 for one way to sidestep this perhaps shocking conclusion).

Challenge Problem 1.2.1: The older brother of Euler's mentor in Basel (Johann Bernoulli) was Jacob Bernoulli (1654–1705), also a talented mathematician. He was highly skilled in summing infinite series,[5] but the problem of $\zeta(2)$ utterly defeated him. When Johann learned of his former student's success, he wrote "If only my brother were alive!" Jacob did have his successes, however. For example, three interesting series he evaluated are:

---

5. Johann was fascinated by infinite series, too. The mysterious integral $\int_0^1 x^x dx$ was done by him in 1697, when he showed the answer is $1 - \frac{1}{2^2} + \frac{1}{3^3} - \frac{1}{4^4} + \frac{1}{5^5} - \cdots = 0.7834\ldots$. This result, called by Bernoulli his "series mirabili" ("marvelous series")—as well as the perhaps even more intimidating $\int_0^1 x^{x^2} dx = 1 - \frac{1}{3^3} + \frac{1}{5^5} - \frac{1}{7^7} + \frac{1}{9^9} - \cdots = 0.8964\ldots$, or its "twin" $\int_0^1 x^{\sqrt{x}} dx = 1 - \left(\frac{2}{3}\right)^2 + \left(\frac{2}{4}\right)^3 - \left(\frac{2}{5}\right)^4 + \left(\frac{2}{6}\right)^5 - \cdots = 0.6585\ldots$—is derived in my *Inside Interesting Integrals* (2nd edition, Springer 2020, pp. 227–229).

$$\sum_{k=1}^{\infty}\frac{k}{2^k}=2, \quad \sum_{k=1}^{\infty}\frac{k^2}{2^k}=6, \quad \sum_{k=1}^{\infty}\frac{k^3}{2^k}=26.$$

Can you discover a general way to sum series like these? If so, confirm Jacob's results, and then do the next obvious sum: $\sum_{k=1}^{\infty}\frac{k^4}{2^k}=?$ Hint: Try differentiating a certain geometric series.

1.3 Euler's Constant, the Zeta Function, and Primes

In 1731 Euler made a curious observation. Writing the harmonic series, (1.2.3), we have

$$(1.3.1) \qquad h(q)=\sum_{k=1}^{q}\frac{1}{k}=\sum_{k=1}^{q-1}\frac{1}{k}+\frac{1}{q}=h(q-1)+\frac{1}{q}.$$

If $h(q)$ were a *continuous* function of q (which it isn't, but just suppose), then we could write

$$\frac{dh}{dq}=\lim_{\Delta q\to 0}\frac{h(q)-h(q-\Delta q)}{\Delta q}.$$

Of course, we are stuck with $\Delta q = 1$, but suppose we ignore that and, using (1.3.1), we write

$$h(q)-h(q-1)=\frac{1}{q}$$

and so

$$(1.3.2) \qquad\qquad \frac{dh}{dq}\approx\frac{1}{q}$$

and argue that (1.3.2) gets "better and better" as $q\to\infty$.

This is all very casual, of course, but in fact it is fairly typical of how Euler found inspiration. We then integrate (1.3.2) indefinitely to get

$$h(q) = \ln(q) + C$$

where C is the constant of indefinite integration. Well, you ask, what *is* C? For Euler,

(1.3.3) $C = \lim_{q \to \infty}\{h(q) - \ln(q)\}.$

Now, as Oresme showed, $h(q)$ blows up in (1.2.4) as $q \to \infty$, but so does $\ln(q)$, and so (perhaps pondered Euler) might their difference approach a finite limit? This is what in fact happens, and C (now usually written as the Greek gamma, γ) has become famous in mathematics as *Euler's constant*. After π and e, γ is perhaps the most important constant in mathematics. We can get an idea of the value of γ by simply plotting (1.3.3), and this is done in the semi-log plot of Figure 1.3.1 as q varies from 1 to 10,000. The plot is certainly not a proof that there is such a limit (maybe for physicists or engineers,

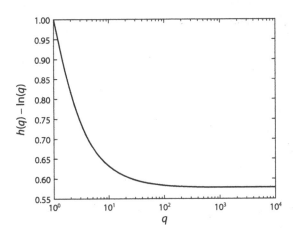

FIGURE 1.3.1.

Computer determination of Euler's constant as a limit.

but not for rigorous mathematicians), but it does strongly suggest that $\gamma \approx 0.57$.

To get our hands on the actual value of γ, we need an analytical expression, and here's one possible way to do that, using the power series expansion for $\ln(1 + x)$, where x is in the interval -1 to 1. This expression was derived by the Danish mathematician Nikolaus Mercator (1620–1687) in 1668, and we'll do it here as follows. We start by observing that

$$(1.3.4) \qquad \frac{1}{1+x} = 1 - x + x^2 - x^3 + x^4 - \cdots,$$

which you can confirm by either doing the long division or by doing the multiplication $(1 + x)(1 - x + x^2 - x^3 + x^4 - \dots)$ and seeing that all the terms cancel except for the leading 1. Then, integrating, (1.3.4) term-by-term yields

$$\int \frac{1}{1+x}\,dx = \ln(1+x) = x - \frac{1}{2}x^2 + \frac{1}{3}x^3 - \frac{1}{4}x^4 + \frac{1}{5}x^5 - \cdots + K$$

where K is the constant of indefinite integration. Setting $x = 0$ gives $\ln(1) = 0 = K$ and so

$$\ln(1+x) = x - \frac{1}{2}x^2 + \frac{1}{3}x^3 - \frac{1}{4}x^4 + \frac{1}{5}x^5 - \cdots$$

or, rearranging,

$$(1.3.5) \qquad x = \ln(1+x) + \frac{1}{2}x^2 - \frac{1}{3}x^3 + \frac{1}{4}x^4 - \frac{1}{5}x^5 + \cdots.$$

Now, successively substitute the values of $x = 1, \frac{1}{2}, \frac{1}{3}, \frac{1}{4}, \cdots, \frac{1}{q}$ into (1.3.5), which gives the following sequence of expressions:

$$1 = \ln(2) + \frac{1}{2} - \frac{1}{3} + \frac{1}{4} - \frac{1}{5} + \cdots$$

$$\frac{1}{2} = \ln\left(\frac{3}{2}\right) + \frac{1}{2}\left(\frac{1}{2^2}\right) - \frac{1}{3}\left(\frac{1}{2^3}\right) + \frac{1}{4}\left(\frac{1}{2^4}\right) - \frac{1}{5}\left(\frac{1}{2^5}\right) + \cdots$$

$$\frac{1}{3} = \ln\left(\frac{4}{3}\right) + \frac{1}{2}\left(\frac{1}{3^2}\right) - \frac{1}{3}\left(\frac{1}{3^3}\right) + \frac{1}{4}\left(\frac{1}{3^4}\right) - \frac{1}{5}\left(\frac{1}{3^5}\right) + \cdots$$

$$\frac{1}{4} = \ln\left(\frac{5}{4}\right) + \frac{1}{2}\left(\frac{1}{4^2}\right) - \frac{1}{3}\left(\frac{1}{4^3}\right) + \frac{1}{4}\left(\frac{1}{4^4}\right) - \frac{1}{5}\left(\frac{1}{4^5}\right) + \cdots$$

$$\cdots$$

$$\frac{1}{q} = \ln\left(\frac{q+1}{q}\right) + \frac{1}{2}\left(\frac{1}{q^2}\right) - \frac{1}{3}\left(\frac{1}{q^3}\right) + \frac{1}{4}\left(\frac{1}{q^4}\right) - \frac{1}{5}\left(\frac{1}{q^5}\right) + \cdots.$$

What do we do with all of these expressions? There are a lot of things we *could* do, but let's do the simplest thing and simply add them. On the left, we immediately get

$$1 + \frac{1}{2} + \frac{1}{3} + \frac{1}{4} + \cdots + \frac{1}{q} = h(q).$$

Next, we'll add the right-hand sides of the expressions in two steps. First, adding all the logarithmic terms gives us

$$\ln(2) + \ln\left(\frac{3}{2}\right) + \ln\left(\frac{4}{3}\right) + \ln\left(\frac{5}{4}\right) + \cdots + \ln\left(\frac{q+1}{q}\right)$$
$$= \ln(2) + \{\ln(3) - \ln(2)\} + \{\ln(4) - \ln(3)\}$$
$$+ \{\ln(5) - \ln(4)\} + \cdots + \{\ln(q+1) - \ln(q)\} = \ln(q+1),$$

because all the terms but the penultimate one cancel (as mathematicians put it, the series *telescopes*). Next, adding the rest of the terms on the right-hand sides of the sequences of expressions together in the highly suggestive way they present themselves (in columns), we see that

$$h(q) - \ln(q+1) = \frac{1}{2}\left\{1 + \frac{1}{2^2} + \frac{1}{3^2} + \frac{1}{4^2} + \cdots + \frac{1}{q^2}\right\} - \frac{1}{3}\left\{1 + \frac{1}{2^3} + \frac{1}{3^3} + \frac{1}{4^3} + \cdots + \frac{1}{q^3}\right\}$$
$$+ \frac{1}{4}\left\{1 + \frac{1}{2^4} + \frac{1}{3^4} + \frac{1}{4^4} + \cdots + \frac{1}{q^4}\right\} - \frac{1}{5}\left\{1 + \frac{1}{2^5} + \frac{1}{3^5} + \frac{1}{4^5} + \cdots + \frac{1}{q^5}\right\} + \cdots$$

or, from (1.3.3)—with its $\ln(q)$ term replaced with $\ln(q + 1)$, which hardly matters, since we are about to let $q \to \infty$ anyway—we have (writing γ now, instead of C)

$$\gamma = \lim_{q \to \infty} \{h(q) - \ln(q+1)\} = \sum_{s=2}^{\infty} \frac{(-1)^s}{s} \left\{ \sum_{q=1}^{\infty} \frac{1}{q^s} \right\}$$

or, amazingly and seemingly out of nowhere,

(1.3.6)
$$\gamma = \sum_{s=2}^{\infty} \frac{(-1)^s}{s} \zeta(s).$$

The intimate connection between Euler's constant and the zeta function is on full display in (1.3.6) but, alas, the sum doesn't converge very rapidly.[6] Still, using just the first 10 terms gives $\gamma \approx 0.5338$ (using the first 100 terms gives $\gamma \approx 0.5723$), which is consistent with Figure 1.3.1. (The actual value is $\gamma = 0.5772156649\ldots$) There is still a lot of mystery to γ. Euler was able to correctly calculate the first few digits (an impressive feat, in its own right), and electronic computers have extended that out to millions of digits. Despite all that, however, it is still not known if γ is rational or not, although every mathematician in the Solar System would be astonished if it turned out to be rational. If it is rational, then it is known that the denominator integer b in $\gamma = a/b$ would have to have hundreds of thousands of digits! As a practical matter, the fraction $228/395$ correctly gives the first six digits, which is almost certainly (as engineers like to put it) "good enough for government work."

---

6. The ultimate convergence of (1.3.6) is, however, guaranteed by the following beautiful little theorem from first-year calculus: an *alternating* series in which the successive terms continually decrease in magnitude toward zero always converges. The issue of the rapidity of the decrease no longer appears. Since (1.2.6) tells us that $\lim_{s \to \infty} \zeta(s) = 1$, then from (1.3.6), we see that $\frac{\zeta(s)}{s} \to 0$ as $s \to \infty$, and so the theorem's requirements are satisfied. At the end of this chapter, I'll show you a generalization of (1.3.6) that converges *much* faster.

Now, for just a moment, let me indulge in a little aside. If you plug $x = 1$ into the power series expansion for $\ln(1 + x)$ you get

$$\ln(2) = 1 - \frac{1}{2} + \frac{1}{3} - \frac{1}{4} + \frac{1}{5} - \frac{1}{6} + \frac{1}{7} - \cdots.$$

That is, if we write the harmonic series with alternating signs as on the right-hand side, the sum now converges, just as footnote 6 claims. There is, however, a subtle, perplexing issue with the convergence of the series: if we sum the terms in a different order, we'll get a different sum. For example, suppose we start with the 1, then add the next two negative terms, then the first skipped positive term, then the next two negative terms, then the next skipped positive term, and so on. Thus, we add the same terms that appear in the $\ln(2)$ expression, but now in the following order:

$$1 - \frac{1}{2} - \frac{1}{4} + \frac{1}{3} - \frac{1}{6} - \frac{1}{8} + \frac{1}{5} - \frac{1}{10} - \frac{1}{12} + \frac{1}{7} - \frac{1}{14} - \frac{1}{16} + \frac{1}{9} - \cdots.$$

If we group these terms as

$$\left(1 - \frac{1}{2}\right) - \frac{1}{4} + \left(\frac{1}{3} - \frac{1}{6}\right) - \frac{1}{8} + \left(\frac{1}{5} - \frac{1}{10}\right) - \frac{1}{12} + \left(\frac{1}{7} - \frac{1}{14}\right) - \frac{1}{16} + \cdots$$

that gives

$$\frac{1}{2} - \frac{1}{4} + \frac{1}{6} - \frac{1}{8} + \frac{1}{10} - \frac{1}{12} + \frac{1}{14} - \cdots = \frac{1}{2}\left(1 - \frac{1}{2} + \frac{1}{3} - \frac{1}{4} + \frac{1}{5} - \frac{1}{6} + \frac{1}{7} - \cdots\right)$$

$$= \frac{1}{2}\ln(2) \neq \ln(2).$$

In 1837 the German mathematician Gustav Dirichlet (1805–1859) proved that for any rearrangement of a series to always converge to the same value, the series must be what mathematicians call *absolutely convergent*. That is, the series must converge even if all its terms are taken as positive (and, as Oresme showed, that is *not* the case for the harmonic series, and that's why we see this curious behavior). In 1854, Riemann observed that the terms of the harmonic series with alternating signs can always be rearranged to converge to any value, positive or negative, that you wish! (For a sketch on how to prove that, see the solution to Challenge Problem 1.3.1.)

In 1737 Euler did something with $\zeta(s)$ that, in some ways, might be even more astounding than is (1.3.6). What he did was show that there is an intimate connection between $\zeta(s)$, a continuous function of s, and the primes (which as integers are the very signature of discontinuity). To start, multiply through (1.2.6), the definition of $\zeta(s)$, by $\frac{1}{2^s}$ to get

$$(1.3.7) \qquad \frac{1}{2^s}\zeta(s) = \frac{1}{2^s} + \frac{1}{4^s} + \frac{1}{6^s} + \frac{1}{8^s} + \frac{1}{10^s} + \frac{1}{12^s} + \cdots,$$

then subtract (1.3.7) from (1.2.6) to arrive at

$$(1.3.8) \qquad \zeta(s) - \frac{1}{2^s}\zeta(s) = \left(1 - \frac{1}{2^s}\right)\zeta(s) = 1 + \frac{1}{3^s} + \frac{1}{5^s} + \frac{1}{7^s}$$

$$+ \frac{1}{9^s} + \frac{1}{11^s} + \frac{1}{13^s} + \cdots.$$

Now, multiply (1.3.8) by $\frac{1}{3^s}$ to get

$$(1.3.9) \qquad \left(1 - \frac{1}{2^s}\right)\frac{1}{3^s}\zeta(s) = \frac{1}{3^s} + \frac{1}{9^s} + \frac{1}{15^s} + \frac{1}{21^s} + \cdots$$

and so, if we subtract (1.3.9) from (1.3.8), we have

$$(1.3.10) \qquad \left(1 - \frac{1}{2^s}\right)\zeta(s) - \left(1 - \frac{1}{2^s}\right)\frac{1}{3^s}\zeta(s) = \left(1 - \frac{1}{2^s}\right)\left(1 - \frac{1}{3^s}\right)\zeta(s)$$

$$= 1 + \frac{1}{5^s} + \frac{1}{7^s} + \frac{1}{11^s} + \cdots.$$

Next, multiply (1.3.10) by $\frac{1}{5^s}$ to get ... and on and on we go, and I'm sure you see the pattern. As we repeat this process over and over, multiplying through our last result by $1/p^s$, where p denotes successive primes, we relentlessly subtract out all the multiples of the primes. You may recognize what we're doing here as essentially executing the famous method called *Eratosthenes' sieve*, developed by the third century BC Greek mathematician Eratosthenes of Cyrene as the fundamental algorithmic procedure for finding all of the primes in the first place.

If we imagine doing this multiply-and-subtract process for all primes, then when we are done (after an infinity of such operations), we will have removed every term but the leading 1 on the right-hand side of (1.3.10) exactly once because of the unique factorization theorem (see note 3). Thus, using Π to denote a product, Euler arrived at

$$\left\{ \prod_{p\ prime} \left(1 - \frac{1}{p^s} \right) \right\} \zeta(s) = 1$$

or, as it is more commonly written,

$$(1.3.11) \qquad \zeta(s) = \prod_{p\ prime} \left(1 - \frac{1}{p^s} \right)^{-1},$$

which is called the Eulerian product form of the zeta function.

In addition to simply being a beautiful expression as it stands, there are two astonishing implications hidden in (1.3.11). One was already known (the infinity of the primes), while the other was new and totally unexpected. To see how Euler had found a new proof for the infinity of the primes, simply notice that if we set $s = 1$, then $\zeta(1)$ is the divergent harmonic series. That is,

$$(1.3.12) \qquad \prod_{p\ prime} \left(1 - \frac{1}{p} \right)^{-1} = \infty.$$

Now, since $p \geq 2$, then every factor in the product is greater than 1, and so the product increases with each additional factor. To increase without bound (that is, to diverge), however, requires that there be an infinite number of factors (that is, an infinity of primes).

This is nice (it's always good to have multiple proofs of a theorem), but it really can't compare with the second, new result Euler extracted from (1.3.11): The sum of the reciprocals of *just the primes, alone*, diverges! It was, after all, a huge surprise when it was realized that the harmonic series, the sum of the reciprocals of all the positive integers, diverges, but to still have divergence even when just the primes are used seems completely and totally unbelievable. Here's how Euler showed, despite that skepticism, that we nevertheless have to believe it. Taking the natural logarithm of (1.3.12), we have (because the log of a product is the sum of the logs, and because $\ln(\infty) = \infty$)

$$(1.3.13) \qquad -\sum_{p \; prime} \ln\left(1 - \frac{1}{p}\right) = \infty.$$

Next, looking back at (1.3.5), if we set $x = -\frac{1}{p}$ then

$$(1.3.14) \quad \ln\left(1 - \frac{1}{p}\right) = -\frac{1}{p} - \frac{1}{2}\left(\frac{1}{p^2}\right) - \frac{1}{3}\left(\frac{1}{p^3}\right) - \frac{1}{4}\left(\frac{1}{p^4}\right) - \frac{1}{5}\left(\frac{1}{p^5}\right) - \cdots$$

and so (1.3.13) becomes

$$\sum_{p \; prime} \left\{\frac{1}{p} + \frac{1}{2}\left(\frac{1}{p^2}\right) + \frac{1}{3}\left(\frac{1}{p^3}\right) + \frac{1}{4}\left(\frac{1}{p^4}\right) + \frac{1}{5}\left(\frac{1}{p^5}\right) + \cdots\right\} = \infty$$

or

$$(1.3.15) \qquad \sum_{p \; prime} \frac{1}{p} + \sum_{p \; prime} \left\{\frac{1}{2}\left(\frac{1}{p^2}\right) + \frac{1}{3}\left(\frac{1}{p^3}\right)\right.$$

$$\left. + \frac{1}{4}\left(\frac{1}{p^4}\right) + \frac{1}{5}\left(\frac{1}{p^5}\right) + \cdots\right\} = \infty.$$

In the second sum on the left, replace every term with a larger one and, in addition, include terms for *every* p (not just for p a prime). Then it is certainly true that

$$\sum_{p \text{ prime}} \left\{ \frac{1}{2}\left(\frac{1}{p^2}\right) + \frac{1}{3}\left(\frac{1}{p^3}\right) + \frac{1}{4}\left(\frac{1}{p^4}\right) + \frac{1}{5}\left(\frac{1}{p^5}\right) + \cdots \right\}$$

$$< \sum_{p=2}^{\infty} \left\{ \left(\frac{1}{p^2}\right) + \left(\frac{1}{p^3}\right) + \left(\frac{1}{p^4}\right) + \left(\frac{1}{p^5}\right) + \cdots \right\}.$$

The expression in the curly brackets on the right is a geometric series, easily summed to give

$$\sum_{p \text{ prime}} \left\{ \frac{1}{2}\left(\frac{1}{p^2}\right) + \frac{1}{3}\left(\frac{1}{p^3}\right) + \frac{1}{4}\left(\frac{1}{p^4}\right) + \frac{1}{5}\left(\frac{1}{p^5}\right) + \cdots \right\} < \sum_{p=2}^{\infty} \frac{1}{p(p-1)}.$$

The sum on the right is easily evaluated, because it telescopes as

$$\sum_{p=2}^{\infty} \frac{1}{p(p-1)} = \sum_{p=2}^{\infty} \left\{ \frac{1}{p-1} - \frac{1}{p} \right\} = \left(1 + \frac{1}{2} + \frac{1}{3} + \frac{1}{4} + \cdots \right)$$

$$- \left(\frac{1}{2} + \frac{1}{3} + \frac{1}{4} + \cdots \right) = 1$$

and so (1.3.15) becomes

$$\sum_{p \text{ prime}} \frac{1}{p} + \left(\text{something less than } 1 \right) = \infty$$

or, just like that,

$$\sum_{p \text{ prime}} \frac{1}{p} = \infty$$

and so the sum of the reciprocals of nothing but the primes, *alone*, diverges.

That is hard to believe, without a doubt, but it's true. As you won't be surprised to learn, the divergence is excruciatingly slow. We know from (1.3.3) that the harmonic series diverges logarithmically: that is, for large q, $h(q) \approx \ln(q)$, where we ignore the "correction" term of γ, which becomes ever-less significant as q increases. The log function is a slowly increasing function of its argument, and so the obvious question now is: What grows even more slowly than the log? I won't prove it here, but an answer is the *iterated-log*, that is, the log of a log. The divergence of the sum of the reciprocals of the primes is as $\ln\{\ln(q)\}$. How good is this estimate? By actual calculation, when the reciprocals of all the primes in the first 1 million integers are added, the result is slightly less than 2.9. The iterated-log estimate gives us

$$\ln\{\ln(10^6)\} = \ln\{6\ln(10)\} = \ln(13.815) = 2.6,$$

which is, in fact, actually not that far off the mark.

Challenge Problem 1.3.1: Write the harmonic series with alternating signs as, first, the sum of all the positive terms, added to the sum of all the negative terms. That is, as

$$\left\{1+\frac{1}{3}+\frac{1}{5}+\frac{1}{7}+\cdots\right\}+\left[-\frac{1}{2}-\frac{1}{4}-\frac{1}{6}+\cdots\right]=A+B.$$

Explain why $A=1+\frac{1}{3}+\frac{1}{5}+\frac{1}{7}+\cdots$ diverges to plus infinity, while $B=-\frac{1}{2}-\frac{1}{4}-\frac{1}{6}+\cdots$ diverges to minus infinity. (Hint: With what you've read in the text, not much more actual math is needed to explain either of these divergences.) Can you use these two conclusions to justify Riemann's observation that there is always some rearrangement of the terms in A and B that will result in the convergence of $A + B$ to *any* value, negative or positive, that you wish?

1.4 Euler's Gamma Function, the Reflection Formula, and the Zeta Function

As 1729 turned to 1730, Euler started the development of what we today call the *gamma function*.[7] This involves a study of the integral

$$(1.4.1) \qquad \Gamma(n) = \int_0^\infty e^{-x} x^{n-1} dx, \ n > 0,$$

which has the wonderful property of extending the idea of the factorial function from just the non-negative integers to all real numbers. Here's how that works.

For $n = 1$, it is easy to calculate

$$(1.4.2) \qquad \Gamma(1) = \int_0^\infty e^{-x} dx = \{-e^{-x}\}\big|_0^\infty = 1.$$

If you integrate by parts, (1.4.1) quickly becomes[8]

$$(1.4.3) \qquad \Gamma(n+1) = n\Gamma(n)$$

and so, for n a positive integer, we immediately see the connection between $\Gamma(n)$ and the factorial function:

$$\Gamma(2) = 1\Gamma(1) = 1(1) = 1!$$

$$\Gamma(3) = 2\Gamma(2) = 2(1!) = 2!$$

$$\Gamma(4) = 3\Gamma(3) = 3(2!) = 3!$$

and so on, all the way to

$$(1.4.4) \qquad \Gamma(n) = (n-1)(n-2)! = (n-1)!$$

7. For an erudite presentation, see Philip J. Davis, "Leonhard Euler's Integral: A Historical Profile of the Gamma Function," *American Mathematical Monthly*, December 1959, pp. 849–869.

8. In $\int_0^\infty u\,dv = \{uv\}\big|_0^\infty - \int_0^\infty v\,du$, let $u = e^{-x}$ and $dv = x^{n-1}\,dx$. Expression (1.4.3) is called the *functional equation* of the gamma function.

Notice, in particular, that setting $n = 1$ in (1.4.4) results in

$$\Gamma(1) = 0!$$

but (1.4.2) then tells us that

$$0! = 1$$

not 0, as most students initially think.[9] To be really emphatic about this,

$$0! \neq 0 \ (!!!!!)$$

In the integral definition of $\Gamma(n)$, in (1.4.1), n does not have to be a positive integer. Indeed, it was the question of how to interpolate the factorial function (for example, $(\frac{1}{2})! = ?$) that motivated Euler to develop the integral definition in the first place. We can also use (1.4.4) to extend n to all real n, including negative values, and so give meaning to objects as strange looking as $(-\frac{1}{2})!$ probably strikes you. Here's how to do that.

First, a specific example. Setting $n = \frac{1}{2}$ in both (1.4.1) and (1.4.4), we have

(1.4.5) $$\Gamma\left(\frac{1}{2}\right) = \int_0^\infty \frac{e^{-x}}{\sqrt{x}} dx = \left(-\frac{1}{2}\right)!$$

Next, change variable to $x = t^2$, and so $dx = 2t \, dt$. Thus,

$$\left(-\frac{1}{2}\right)! = \int_0^\infty \frac{e^{-t^2}}{t} 2t \, dt = 2\int_0^\infty e^{-t^2} dt = 2\left\{\frac{1}{2}\sqrt{\pi}\right\}$$

---

9. A more direct way to arrive at this result is to write $n! = n(n-1)!$ and then set $n = 1$. Thus, $1! = 1(0!) = 0!$ Since $1! = 1$ we have, again, $0! = 1$.

and so

$$(1.4.6) \qquad \Gamma\left(\frac{1}{2}\right) = \left(-\frac{1}{2}\right)! = \sqrt{\pi}.$$

Who would have believed that $(-\frac{1}{2})!$ could mean *anything*, before Euler came along? Until Euler, nobody had even wondered about such a weird thing.[10] (That last integral, $\int_0^\infty e^{-t^2} dt$, of course, needs some explaining; see Appendix 2 for a derivation.)

Now, let's be more general. If, in (1.4.4), we replace n with $1 - n$ on both sides of the expression, we obtain the interesting result

$$(1.4.7) \qquad \Gamma(1-n) = (1-n-1)! = (-n)!$$

Since from (1.4.4) we have

$$n\Gamma(n) = n(n-1)! = n!$$

then

$$(1.4.8) \qquad n\Gamma(n)\Gamma(1-n) = (n!)(-n)!$$

That is, if we could evaluate the left-hand side of (1.4.8), we would then have a way to calculate $(-n)!$ from the value of $n!$, for any $n \geq 0$. An evaluation of $n\Gamma(n)\Gamma(1-n)$ can, in fact, be done by working directly with the integral definition of the gamma function, (1.4.1), but that approach has (for us, in this book) the drawback of using some mathematics that is just beyond AP-calculus.[11] What I'll

---

10. See if you can calculate $(\frac{1}{2})!$ right now. I'll ask you to think about this again a little later in this section and, yet again, at the end of this section as a challenge question.

11. See my *An Imaginary Tale: The Story of $\sqrt{-1}$* (Princeton University Press, 2016), pp. 182–184. That discussion concludes with an evaluation of the integral $\int_0^\infty \frac{s^{a-1}}{1+s} ds$, which is done using complex function theory (contour integration), a topic developed in that book on its pp. 187–226.

show you next, instead, is a way to calculate $(n!)(-n)!$ which neatly avoids that problem.

To start, let's develop another of Euler's beautiful discoveries, one that we'll need in just a bit. This is his formulation of the sine function as an infinite product:

$$(1.4.9) \qquad \sin(y) = y \prod_{n=1}^{\infty} \left(1 - \frac{y^2}{n^2 \pi^2} \right).$$

This famous expression is normally established with some pretty sophisticated mathematics, but I'll limit my comments here to a series of plausible assertions (but I think you'll find them pretty convincing). If we write $\sin(y)$ as a power series, that is, as

$$\sin(y) = y - \frac{1}{3!} y^3 + \frac{1}{5!} y^5 - \cdots,$$

or, dividing through by y,

$$(1.4.10) \qquad \frac{\sin(y)}{y} = 1 - \frac{1}{3!} y^2 + \frac{1}{5!} y^4 - \cdots,$$

then it doesn't seem unreasonable to say (as, in fact, did Euler) that $\frac{\sin(y)}{y}$ is a *polynomial of infinite degree*. Notice, in particular, that y appears in (1.4.10) raised only to ever-increasing even powers.

Now, fall back on your algebraic experience with polynomials of finite degree. If somebody told you she was thinking of a polynomial $P(y)$ of degree s, with non-zero roots $r_1, r_2, \ldots, r_s$, then to within a scale factor of A you'd write that polynomial as the product

$$P(y) = A(y - r_1)(y - r_2) \cdots (y - r_s).$$

If we write each factor as $y - r_k = -(r_k - y)$ and absorb the s minus signs into A, then

$$P(y) = A(r_1 - y)(r_2 - y) \cdots (r_s - y)$$

or, factoring out $r_1, r_2, \ldots, r_s$,

$$P(y) = A\left[r_1\left(1-\frac{y}{r_1}\right)r_2\left(1-\frac{y}{r_2}\right)\cdots r_s\left(1-\frac{y}{r_s}\right)\right].$$

Again, if we absorb the product $r_1 r_2 \ldots r_s$ into A,

(1.4.11) $$P(y) = A\left[\left(1-\frac{y}{r_1}\right)\left(1-\frac{y}{r_2}\right)\cdots\left(1-\frac{y}{r_s}\right)\right].$$

We know the roots of $\sin(y) = 0$ are y equal to any integer multiple of π. That is, the roots are $y = 0, \pm\pi, \pm2\pi, \pm3\pi$, and so on, or equivalently, $y^2 = 0, \pi^2, 2^2\pi^2, 3^2\pi^2$, and so on. The situation for $\frac{\sin(y)}{y}$ is the same, with the exception that $y = 0$ is *not* a root of $\frac{\sin(y)}{y} = 0$, because $\lim_{y\to 0}\frac{\sin(y)}{y} = 1$ by L'Hôpital's rule. So, using (1.4.11) as a guide, it seems reasonable to jump from the finite to the infinite (but, to be honest, this is not always legitimate!) and write

(1.4.12) $$\frac{\sin(y)}{y} = A\left[\left(1-\frac{y^2}{\pi^2}\right)\left(1-\frac{y^2}{2^2\pi^2}\right)\left(1-\frac{y^2}{3^2\pi^2}\right)\cdots\right].$$

Notice that (1.4.12) has y raised only to *even* powers of y, just as in (1.4.10). Since the left-hand side of (1.4.12) is 1 at $y = 0$, and the right-hand side is A, we have

$$\frac{\sin(y)}{y} = \prod_{n=1}^{\infty}\left(1-\frac{y^2}{n^2\pi^2}\right)$$

or

$$\sin(y) = y\prod_{n=1}^{\infty}\left(1-\frac{y^2}{n^2\pi^2}\right),$$

which is (1.4.9). (See the following box for how the value for $\zeta(2)$ follows almost immediately from (1.4.9).)

From (1.4.9), and writing $\sin(y)$ as a power series, we have

$$y - \frac{1}{3!}y^3 + \frac{1}{5!}y^5 - \cdots = y\left(1 - \frac{y^2}{\pi^2}\right)\left(1 - \frac{y^2}{2^2\pi^2}\right)\left(1 - \frac{y^2}{3^2\pi^2}\right)\cdots$$

$$= y - \left(\frac{1}{\pi^2} + \frac{1}{2^2\pi^2} + \frac{1}{3^2\pi^2} + \cdots\right)y^3 + \text{higher-order terms.}$$

Equating the coefficients of the y^3 term on each side,

$$-\frac{1}{3!} = -\left(\frac{1}{\pi^2} + \frac{1}{2^2\pi^2} + \frac{1}{3^2\pi^2} + \cdots\right)$$

or,

$$\frac{\pi^2}{3!} = \frac{\pi^2}{6} = 1 + \frac{1}{2^2} + \frac{1}{3^2} + \cdots = \zeta(2).$$

Done!

Okay, back to calculating $(n!)(-n)!$

Let's initially assume n is a positive integer, and so

(1.4.13) $n! = 1 \cdot 2 \cdot 3 \cdots (n-1)n$

and then we'll manipulate (1.4.13)—using a method due to the German mathematician Karl Weierstrass (1815–1897)—until we get an expression that makes sense even when n is not an integer, or even positive. So, suppose a is another integer whose particular value doesn't matter, because we are going to be taking the limit $a \to \infty$. Multiplying (1.4.13) by 1, we have

$$n! = \lim_{\alpha \to \infty}\left[\{1 \cdot 2 \cdot 3 \cdots (n-1)n\} \frac{(n+1)(n+2)(n+3)\cdots(n+\alpha)\alpha^n}{(n+1)(n+2)(n+3)\cdots(n+\alpha)\alpha^n} \right].$$

Now, as $\alpha \to \infty$, it is safe to assume that, at some point, α will exceed n and so allow us to write

$$n! = \lim_{\alpha \to \infty}\left[\frac{\{1 \cdot 2 \cdot 3 \cdots (n-1)n(n+1)(n+2)\cdots\alpha\}\{(\alpha+1)(\alpha+2)\cdots(\alpha+n)\}\alpha^n}{\{(n+1)(n+2)(n+3)\cdots(n+\alpha)\}\alpha^n} \right]$$

$$= \lim_{\alpha \to \infty}\left[\left\{ \frac{\alpha!\alpha^n}{(n+1)(n+2)(n+3)\cdots(n+\alpha)} \right\} \left\{ \frac{(\alpha+1)(\alpha+2)\cdots(\alpha+n)}{\alpha^n} \right\} \right].$$

As $\alpha \to \infty$, for any finite value of n, we can (with vanishing error) replace the product $(\alpha + 1)(\alpha + 2) \ldots (\alpha + n)$ with α^n in the numerator of the right-most curly brackets, and so we have

$$n! = \lim_{\alpha \to \infty} \frac{\alpha!\alpha^n}{(n+1)(n+2)(n+3)\cdots(n+\alpha)}$$

$$= \lim_{\alpha \to \infty} \frac{\alpha!\alpha^n}{\left(1+\frac{n}{1}\right)2\left(1+\frac{n}{2}\right)3\left(1+\frac{n}{3}\right)\cdots\alpha\left(1+\frac{n}{\alpha}\right)}$$

$$= \lim_{\alpha \to \infty} \frac{\alpha!\alpha^n}{\alpha!\left(1+\frac{n}{1}\right)\left(1+\frac{n}{2}\right)\left(1+\frac{n}{3}\right)\cdots\left(1+\frac{n}{\alpha}\right)}$$

or

$$(1.4.14) \qquad n! = \lim_{\alpha \to \infty} \frac{\alpha^n}{\left(1+\frac{n}{1}\right)\left(1+\frac{n}{2}\right)\left(1+\frac{n}{3}\right)\cdots\left(1+\frac{n}{\alpha}\right)}.$$

Next, write

$$\alpha^n = e^{\ln(\alpha^n)} = e^{n\ln(\alpha)}$$

and then recall (1.3.3), which defines Euler's constant γ:

$$\gamma = \lim_{\alpha \to \infty}\{h(\alpha) - \ln(\alpha)\}$$

where $h(\alpha)$ is defined in (1.2.3). That is, as $\alpha \to \infty$, we'll write

$$\ln(\alpha) = -\gamma + \sum_{k=1}^{\alpha}\frac{1}{k} = -\gamma + \left[1 + \frac{1}{2} + \frac{1}{3} + \cdots + \frac{1}{\alpha}\right]$$

and so, as $\alpha \to \infty$,

$$(1.4.15) \qquad \alpha^n = e^{-n\gamma + n\left[1 + \frac{1}{2} + \frac{1}{3} + \ldots + \frac{1}{\alpha}\right]} = e^{-n\gamma}e^n e^{\frac{n}{2}} e^{\frac{n}{3}} \ldots e^{\frac{n}{\alpha}}.$$

Putting (1.4.15) into (1.4.14),

$$n! = \lim_{\alpha \to \infty} \frac{e^{-n\gamma}e^n e^{\frac{n}{2}} e^{\frac{n}{3}} \cdots e^{\frac{n}{\alpha}}}{\left(1 + \frac{n}{1}\right)\left(1 + \frac{n}{2}\right)\left(1 + \frac{n}{3}\right)\cdots\left(1 + \frac{n}{\alpha}\right)}$$

or

$$(1.4.16) \qquad n! = e^{-n\gamma} \lim_{\alpha \to \infty} \frac{e^n}{\left(1 + \frac{n}{1}\right)} \frac{e^{\frac{n}{2}}}{\left(1 + \frac{n}{2}\right)} \frac{e^{\frac{n}{3}}}{\left(1 + \frac{n}{3}\right)} \cdots \frac{e^{\frac{n}{\alpha}}}{\left(1 + \frac{n}{\alpha}\right)}.$$

Replacing n with $-n$ in (1.4.16), we have

$$(1.4.17) \qquad (-n)! = e^{n\gamma} \lim_{\alpha \to \infty} \frac{e^{-n}}{\left(1 - \frac{n}{1}\right)} \frac{e^{-\frac{n}{2}}}{\left(1 - \frac{n}{2}\right)} \frac{e^{-\frac{n}{3}}}{\left(1 - \frac{n}{3}\right)} \cdots \frac{e^{-\frac{n}{\alpha}}}{\left(1 - \frac{n}{\alpha}\right)}$$

and so, if we multiply (1.4.16) and (1.4.17) together, we see lots of mutual annihilations that give us

$$(n!)(-n)! = \lim_{\alpha \to \infty} \frac{1}{\left(1 - \frac{n^2}{1^2}\right)} \frac{1}{\left(1 - \frac{n^2}{2^2}\right)} \frac{1}{\left(1 - \frac{n^2}{3^2}\right)} \cdots \frac{1}{\left(1 - \frac{n^2}{\alpha^2}\right)}$$

or

(1.4.18)
$$(n!)(-n)! = \frac{1}{\prod_{k=1}^{\infty}\left(1 - \frac{n^2}{k^2}\right)}.$$

Does (1.4.18) look familiar? It should! If we rewrite (1.4.9) with a trivial change in the product index from n to k, we have

$$\sin(y) = y \prod_{k=1}^{\infty}\left(1 - \frac{y^2}{k^2 \pi^2}\right)$$

and so, if we set $y = n\pi$, we have

$$\sin(n\pi) = n\pi \prod_{k=1}^{\infty}\left(1 - \frac{n^2}{k^2}\right)$$

or

$$\prod_{k=1}^{\infty}\left(1 - \frac{n^2}{k^2}\right) = \frac{\sin(n\pi)}{n\pi}.$$

Thus, (1.4.18) becomes the beautiful

(1.4.19)
$$(n!)(-n)! = \frac{n\pi}{\sin(n\pi)},$$

which "makes sense," even if n is not an integer.

We can use (1.4.19) to answer the question raised in note 10—$(\frac{1}{2})! = ?$— using the result in (1.4.6) for the value of $(-\frac{1}{2})!$ That is, with $n = \frac{1}{2}$, (1.4.19) says

$$\left(\frac{1}{2}\right)! = \frac{\dfrac{1}{2}\pi}{\left(-\dfrac{1}{2}\right)!\sin\left(\dfrac{1}{2}\pi\right)} = \frac{\dfrac{1}{2}\pi}{\sqrt{\pi}} = \frac{1}{2}\sqrt{\pi}.$$

This is the value of $(\frac{1}{2})!$ *if* the analysis leading up to (1.4.19) is correct. As note 10 hints, we'll return one last time to the question of $(\frac{1}{2})!$ in a challenge problem at the end of this section, where I'll ask you to confirm this result by deriving it via a completely different method.

The validity of (1.4.19) depends on the correctness in (1.4.9) of Euler's infinite product form of the sine function. There is an immediate prediction that follows from (1.4.9) that gives us confidence in its truth (even though we/Euler derived it in a less than rigorous way). If we put $y = \frac{\pi}{2}$ in (1.4.9), we get

$$\sin\left(\frac{\pi}{2}\right) = 1 = \frac{\pi}{2}\left(1 - \frac{1}{4}\right)\left(1 - \frac{1}{16}\right)\left(1 - \frac{1}{36}\right)\cdots$$

$$= \frac{\pi}{2}\left[\left(1 - \frac{1}{2}\right)\left(1 + \frac{1}{2}\right)\right]\left[\left(1 - \frac{1}{4}\right)\left(1 + \frac{1}{4}\right)\right]\left[\left(1 - \frac{1}{6}\right)\left(1 + \frac{1}{6}\right)\right]\cdots$$

$$= \frac{\pi}{2}\left[\left(\frac{1}{2}\right)\left(\frac{3}{2}\right)\right]\left[\left(\frac{3}{4}\right)\left(\frac{5}{4}\right)\right]\left[\left(\frac{5}{6}\right)\left(\frac{7}{6}\right)\right]\cdots.$$

That is,

$$1 = \frac{\pi}{2}\left(\frac{1 \cdot 3}{2 \cdot 2}\right)\left(\frac{3 \cdot 5}{4 \cdot 4}\right)\left(\frac{5 \cdot 7}{6 \cdot 6}\right)\cdots$$

or, as it is usually written,

$$\frac{\pi}{2} = \left(\frac{2 \cdot 2}{1 \cdot 3}\right)\left(\frac{4 \cdot 4}{3 \cdot 5}\right)\left(\frac{6 \cdot 6}{5 \cdot 7}\right)\cdots\left(\frac{2n \cdot 2n}{(2n-1)(2n+1)}\right)\cdots,$$

which is a famous result due to the English mathematician John Wallis (1616–1703), who discovered it in 1655 by entirely different means. The fact that (1.4.9) agrees with Wallis' result is strong evidence that Euler's infinite product is correct. We'll use Wallis' formula in the next chapter to derive an important result concerning the zeta function.

There is an elegant result tucked away in (1.4.8) and (1.4.19) that was discovered by Euler in 1771. Writing those two equations on a single line, we have

$$n\Gamma(n)\Gamma(1-n) = (n!)(-n)! = \frac{n\pi}{\sin(n\pi)}$$

and so, just like that (once we make the obvious cancellations of n's in the first and last terms), we have Euler's famous *reflection formula* for the gamma function:

(1.4.20) $$\Gamma(n)\Gamma(1-n) = \frac{\pi}{\sin(n\pi)}.$$

(Do you see why the word *reflection* is used in the name? Think about what $1 - n$ does as n increases through positive values.)

In 1859 Riemann showed that there is an intimate connection between the gamma and zeta functions. Here's what he did. Start with the integral

$$\int_0^\infty e^{-nx} x^{s-1}\, dx$$

and then change variable to $u = nx$, where n is some (any) positive integer. That is, write

$$\int_0^\infty e^{-nx} x^{s-1}\, dx = \int_0^\infty e^{-u}\left(\frac{u}{n}\right)^{s-1}\frac{du}{n} = \int_0^\infty \frac{e^{-u} u^{s-1}}{n^s}\, du$$

and so, from (1.4.1), the definition of the gamma function, we have

$$(1.4.21) \qquad \int_0^\infty e^{-nx} x^{s-1}\, dx = \frac{\Gamma(s)}{n^s}.$$

Then, summing (1.4.21) over all possible n,

$$(1.4.22) \quad \sum_{n=1}^\infty \int_0^\infty e^{-nx} x^{s-1}\, dx = \sum_{n=1}^\infty \frac{\Gamma(s)}{n^s} = \Gamma(s) \sum_{n=1}^\infty \frac{1}{n^s} = \Gamma(s)\zeta(s).$$

If we assume that we can interchange the order of the summation and integration operations at the far left of (1.4.22), then we have[12]

$$(1.4.23) \qquad \Gamma(s)\zeta(s) = \int_0^\infty x^{s-1} \left\{ \sum_{n=1}^\infty e^{-nx} \right\} dx.$$

The summation in (1.4.23) is a geometric sum and easy to do:

$$\sum_{n=1}^\infty e^{-nx} = e^{-x} + e^{-2x} + e^{3x} + \cdots = \frac{1}{e^x - 1}$$

and (1.4.23) becomes Riemann's famous integral formula:

$$(1.4.24) \qquad \int_0^\infty \frac{x^{s-1}}{e^x - 1}\, dx = \Gamma(s)\zeta(s).$$

For $s = 2$, for example, (1.4.24) says

$$\int_0^\infty \frac{x}{e^x - 1}\, dx = \Gamma(2)\zeta(2) = (1!)\frac{\pi^2}{6} = \frac{\pi^2}{6}$$

12. Reversing the order of integration and summation in an analysis is something most physicists and engineers do without hesitation. It is, however, *not* always legitimate. However, since no small creatures of the forest will be harmed by doing it here (and if we promise to toss our calculations into the trash if we get an obviously dumb answer), let's throw all caution to the wind and do it anyway!

and for $s = 4$,

$$\int_0^\infty \frac{x^3}{e^x - 1} dx = \Gamma(4)\zeta(4) = (3!)\frac{\pi^4}{90} = \frac{\pi^4}{15}.$$

This last integral appears, for example, in the theory of blackbody radiation, and is therefore of great interest to physicists as well as to mathematicians.[13]

And, as one more example, for $s = 3$, we have

$$\int_0^\infty \frac{x^2}{e^x - 1} dx = \Gamma(3)\zeta(3) = (2!)\zeta(3) = 2\zeta(3)$$

or

(1.4.25) $$\zeta(3) = \frac{1}{2}\int_0^\infty \frac{x^2}{e^x - 1} dx,$$

which expresses zeta-3 as an integral rather than a discrete sum. We'll see more of such integral forms for zeta-3 in Chapter 2.

Challenge Problem 1.4.1: Using (1.4.19), we calculated $(\frac{1}{2})! = \frac{1}{2}\sqrt{\pi}$. Redo this calculation as a direct evaluation of the defining integral for the gamma function. That is, set $n = \frac{3}{2}$ in (1.4.1) and (1.4.4), integrate (1.4.4) by parts, and use (1.4.5)/(1.4.6). Do you get the same answer that (1.4.19) gave? Hint: You should!

Challenge Problem 1.4.2: Explain why $|k!|$ is infinity for k any *negative* integer.

1.5 Ramanujan's Master Theorem

In this section, we'll take a brief side trip to develop a beautiful "close cousin" to Riemann's integral formula of (1.4.24), a formula that also

---

13. See, for example, Richard Feynman, *Lectures on Physics*, vol. 1 (Addison-Wesley, 1963), pp. 45–48.

involves the gamma function. What we'll do will strike the careful reader as, at best, perhaps odd, but powerful new mathematics is often first developed by not being bound by "the rules." Afterward, more careful analyses can straighten out the kinks when writing it all up for publication.[14] We start by defining the so-called *forward shift* operator[15] E as follows:

(1.5.1) $E\{\lambda(n)\} = \lambda(n + 1)$.

Thus,

$$E\{\lambda(0)\} = \lambda(1)$$

$$E\{\lambda(1)\} = E\{E\{\lambda(0)\}\} = E^2\{\lambda(0)\} = \lambda(2)$$

$$E\{\lambda(2)\} = E\{E\{E\{\lambda(0)\}\}\} = E^3\{\lambda(0)\} = \lambda(3)$$

and so on. In general,

14. A famous example of this is the eventual acceptance into legitimate mathematics of the *impulse function*, defined as $\delta(x) = \{^{\infty, x=0}_{0, x\neq0}$, with the property $\int_{-\infty}^{\infty} \delta(x)dx = 1$. This function was used in the late 1920s by the English physicist Paul Dirac (1902–1984) to solve longstanding puzzling problems in quantum mechanics, even though mathematicians thought the impulse definition to be absurd (how could a function that is zero everywhere with the exception of one point nevertheless bound a unit area?). Then, decades later, the French mathematician Laurent Schwartz (1915–2002) put $\delta(x)$ on a solid theoretical foundation, for which he received one of mathematics' greatest prizes in 1950, the Fields Medal. (For *his* pioneering work, Dirac had earlier received the 1933 Nobel Prize in physics.)

15. *Operators* are highly useful concepts and are actually quite common in "higher math." A study of differential equations, for example, quickly leads to the differential operator D, defined as $D\{\lambda\} = \frac{d\lambda}{dt}$. The fact that operators can often be manipulated just as if they were numbers makes their use intuitive, and that feature will let us arrive at useful results through the use of only formal manipulations, rather than rigorous reasoning. (The word *formal* is a technical term for "symbol pushing," an activity usually frowned upon by pure mathematicians, but not so much by engineers and physicists.) I'll say a bit more about the differential operator in the next chapter.

(1.5.2) $E^k\{\lambda(0)\} = \lambda(k).$

Next, imagine that we have some function $\varphi(x)$ with the power series expansion

(1.5.3) $\varphi(x) = \sum_{n=0}^{\infty} \dfrac{(-1)^n}{n!}\lambda(n)x^n.$

Then, continuing in a *formal* way (see note 15 again), we write

(1.5.4) $\displaystyle\int_0^{\infty} x^{s-1}\varphi(x)\,dx = \int_0^{\infty} x^{s-1}\sum_{n=0}^{\infty}\dfrac{(-1)^n}{n!}\lambda(n)x^n\,dx.$

Using (1.5.2), we replace $\lambda(n)$ in (1.5.4) with $E^n\{\lambda(0)\}$, which we loosely write (by dropping the curly brackets) as $E^n\lambda(0)$. Continuing with our formal manipulations, then, we have

$$\int_0^{\infty} x^{s-1}\varphi(x)\,dx = \int_0^{\infty} x^{s-1}\sum_{n=0}^{\infty}\dfrac{(-1)^n}{n!}E^n\lambda(0)x^n\,dx$$

$$= \lambda(0)\int_0^{\infty} x^{s-1}\sum_{n=0}^{\infty}\dfrac{(-1)^n}{n!}(Ex)^n\,dx.$$

Well, what can I say: There is simply no denying that we've really played pretty fast-and-loose with the operator E in arriving at the last integral (in particular, notice how E^n goes from operating on $\lambda(0)$ to operating on x^n, while the $\lambda(0)$ slides outside to the front of the integral—this is symbol pushing at its most arrogant). Take a deep breath, however, as there is still *more* of such doings to come!

Recalling the power series expansion of e^{-t}

$$e^{-t} = 1 - t + \dfrac{t^2}{2!} - \dfrac{t^3}{3!} + \cdots = \sum_{n=0}^{\infty}\dfrac{(-1)^n}{n!}t^n$$

and writing $t = Ex$, then we *formally* have

$$(1.5.5) \qquad \int_0^\infty x^{s-1} \varphi(x) dx = \lambda(0) \int_0^\infty x^{s-1} e^{-Ex} dx.$$

Treating E as a number, let's now change variable to $y = Ex$. Then

$$dx = \frac{1}{E} dy$$

and (1.5.5) becomes

$$(1.5.6) \quad \int_0^\infty x^{s-1} \varphi(x) dx = \lambda(0) \int_0^\infty \left(\frac{y}{E}\right)^{s-1} e^{-y} \frac{1}{E} dy = \frac{\lambda(0)}{E^s} \int_0^\infty y^{s-1} e^{-y} dy.$$

From (1.4.1), we see that the right-most integral in (1.5.6) is $\Gamma(s)$ and so, continuing,

$$\int_0^\infty x^{s-1} \varphi(x) dx = \frac{\lambda(0)}{E^s} \Gamma(s) = E^{-s} \lambda(0) \Gamma(s).$$

Finally, recalling (1.5.2) with $k = -s$, we have

$$(1.5.7) \qquad \int_0^\infty x^{s-1} \varphi(x) dx = \Gamma(s) \lambda(-s),$$

a result called *Ramanujan's Master Theorem* (or RMT), named after the Indian mathematician Srinivasa Ramanujan (1887–1920), who discovered (1.5.7) sometime during the first decade of the 20th century.[16]

---

16. For a brief historical discussion of the RMT, and of how it was very nearly discovered decades before Ramanujan (in 1874) by the English mathematician J. W. L. Glaisher (1848–1928), see V. H. Moll et al., "Ramanujan's Master Theorem," *The Ramanujan Journal*, December 2012, pp. 103–120. For more on Ramanujan's discovery of the RMT, see Bruce C. Berndt, "The Quarterly Reports of S. Ramanujan," *American Mathematical Monthly*, October 1983, pp. 505–516. In advanced math, the integral in (1.5.7) is called the *Mellin transform* of $\varphi(x)$, after the Finnish mathematician Robert Hjalmar Mellin (1854–1933). In computer science, the Mellin transform is used in electronic image processing, an application that Mellin and Ramanujan couldn't have even imagined in their wildest dreams.

This development of (1.5.7) almost surely strikes you—as it should—as being pretty bizarre, akin to arriving at the correct statement that $\frac{16}{64} = \frac{1}{4}$ by simply canceling the two 6's! I've purposely taken you through what most mathematicians would call an outrageous "derivation" for two reasons: (1) it closely resembles how Ramanujan himself did it (Euler would have loved it!), and (2) the result in (1.5.7) *can* be rigorously established using mathematics beyond AP-calculus (an approach I wish, however, to avoid in this book). You'll be relieved to learn that when Ramanujan made his famous journey to England in 1914 to come under the tutelage of his hero, the great G. H. Hardy (events dramatically described in Robert Kanigel's 1991 book *The Man Who Knew Infinity*), he learned both that his "derivation" was technically faulty as well as how to do it right.[17]

As an example of the RMT in action, suppose that

$$\varphi(x) = (1 + x)^m.$$

By the binomial theorem,

$$(1+x)^m = 1 + mx + \frac{m(m-1)}{2!}x^2 + \frac{m(m-1)(m-2)}{3!}x^3 + \cdots.$$

If $m = -a$, where $a > 0$, then[18]

---

17. See, for example, G. H. Hardy, *Ramanujan: Twelve Lectures on Subjects Suggested by His Life and Work* (Chelsea, 1978), based on lectures given in 1936 by Hardy at Harvard University.

18. The binomial theorem, for m a positive integer, gives a finite number of terms and has been known since the French mathematician Blaise Pascal (1623–1662) discovered it. The English mathematical physicist Isaac Newton (1642–1727) extended its use (without proof) to rational m (and even negative m), for which the expansion has an infinite number of terms. The proof of the theorem, for any m in general, wasn't done until the Norwegian mathematician Niels Henrik Abel (1802–1829) did it in 1826.

$$(1+x)^{-a} = 1 - ax + \frac{-a(-a-1)}{2!}x^2 + \frac{-a(-a-1)(-a-2)}{3!}x^3 + \cdots$$

$$= 1 + \frac{a}{1!}(-1)^1 x + \frac{(a)(a+1)}{2!}(-1)^2 x^2 + \frac{(a)(a+1)(a+2)}{3!}(-1)^3 x^3 + \cdots.$$

We see that the general term is

$$\frac{(a)(a+1)(a+2)\cdots(a+n-1)}{n!}(-1)^n x^n$$

or, writing the numerator product in reverse order,

$$\frac{(a+n-1)\cdots(a+2)(a+1)(a)}{n!}(-1)^n x^n$$

$$= \frac{(a+n-1)\cdots(a+2)(a+1)(a)(a-1)!}{n!(a-1)!}(-1)^n x^n$$

$$= \frac{(a+n-1)!}{(a-1)!}\frac{(-1)^n}{n!}x^n.$$

That is,

$$(1+x)^{-a} = \sum_{n=0}^{\infty} \frac{(a+n-1)!}{(a-1)!}\frac{(-1)^n}{n!}x^n$$

or, using the gamma function notation of (1.4.4),

$$(1.5.8) \qquad (1+x)^{-a} = \sum_{n=0}^{\infty} \frac{\Gamma(a+n)}{\Gamma(a)}\frac{(-1)^n}{n!}x^n.$$

Comparing the right-hand side of (1.5.8) with the right-hand side of (1.5.3), we see that

$$\lambda(n) = \frac{\Gamma(a+n)}{\Gamma(a)}$$

and so

$$\lambda(-s)=\frac{\Gamma(a-s)}{\Gamma(a)}.$$

The RMT then gives us the very pretty result

(1.5.9) $$\int_0^\infty \frac{x^{s-1}}{(1+x)^a}dx=\frac{\Gamma(s)\Gamma(a-s)}{\Gamma(a)}, a>0.$$

In particular, if we let $a=1$ and $s-1=-p$ (and so $s=1-p$), then

$$\int_0^\infty \frac{x^{-p}}{1+x}dx=\int_0^\infty \frac{1}{(1+x)x^p}dx=\frac{\Gamma(1-p)\Gamma(p)}{\Gamma(1)}$$

or, remembering both the reflection formula and that $\Gamma(1)=1$, we arrive at the following interesting integral (which is generally developed in textbooks as the result of a contour integration in the complex plane):[19]

(1.5.10) $$\int_0^\infty \frac{1}{(1+x)x^p}dx=\frac{\pi}{\sin(p\pi)}, 0<p<1.$$

Challenge Problem 1.5.1: Evaluate $\int_{-\infty}^\infty \frac{e^{px}}{1+e^x}dx$. Hint: Make the appropriate change of variable, and then recall (1.5.10).

1.6 Integral Forms for the Harmonic Series and Euler's Constant

As a prelude to developing integral forms for zeta-3 (which we'll do in the next chapter), I'll now show you how to develop similar integral forms for the harmonic series and Euler's constant. To start, for q a non-negative integer, consider the integral

19. See, for example, my *Inside Interesting Integrals*, 2nd edition (Springer, 2020), pp. 388–392.

(1.6.1) $\int_0^1 \dfrac{1-(1-x)^q}{x}dx.$

Changing variable to $u = 1 - x$ (and so $dx = -du$), we have

$$\int_0^1 \frac{1-(1-x)^q}{x}dx = \int_1^0 \frac{1-u^q}{1-u}(-du)$$

$$= \int_0^1 \frac{(1-u)(1+u+u^2+u^3+\cdots+u^{q-1})}{1-u}du$$

$$= \int_0^1 (1+u+u^2+u^3+\cdots+u^{q-1})du = \left(u+\frac{1}{2}u^2+\frac{1}{3}u^3+\cdots+\frac{1}{q}u^q\right)\Big|_0^1$$

$$= 1+\frac{1}{2}+\frac{1}{3}+\cdots+\frac{1}{q} = h(q),$$

as written in (1.2.3). As an aside, notice that writing "for q a non-negative integer" means $q = 0$ is possible. This creates an apparent puzzle when thinking of the $h(q)$ *series* (which starts at $q = 1$), because the integral in (1.6.1) continues to make sense at $q = 0$ as it says $h(0) = 0$. This is a condition we'll make great use of in Chapter 4, and I mention it now just to get you thinking about the $q = 0$ case.[20]

Returning to the integral in (1.6.1), which we now know is $h(q)$, change the variable to $u = qx$ (and so $dx = \frac{1}{q}du$). For the case of $q \geq 1$ (we are eventually, in fact, going to let $q \to \infty$),

$$h(q)=\int_0^q \frac{1-\left(1-\dfrac{u}{q}\right)^q}{\dfrac{u}{q}}\left(\frac{1}{q}du\right)=\int_0^q \frac{1-\left(1-\dfrac{u}{q}\right)^q}{u}du$$

$$=\int_0^1 \frac{1-\left(1-\dfrac{u}{q}\right)^q}{u}du+\int_1^q \frac{1-\left(1-\dfrac{u}{q}\right)^q}{u}du$$

20. But perhaps there really isn't a puzzle. From (1.2.3), we have $\sum_{k=1}^q \frac{1}{k} = h(q-1)+\frac{1}{q}$, and so if $q = 1$, this says $1 = h(0) + 1$ or, again, $h(0) = 0$.

$$= \int_0^1 \frac{1-\left(1-\dfrac{u}{q}\right)^q}{u} du + \int_1^q \frac{1}{u} du - \int_1^q \frac{\left(1-\dfrac{u}{q}\right)^q}{u} du = \int_0^1 \frac{1-\left(1-\dfrac{u}{q}\right)^q}{u} du + \ln(q) - \int_1^q \frac{\left(1-\dfrac{u}{q}\right)^q}{u} du.$$

So,

$$h(q) - \ln(q) = \int_0^1 \frac{1-\left(1-\dfrac{u}{q}\right)^q}{u} du - \int_1^q \frac{\left(1-\dfrac{u}{q}\right)^q}{u} du.$$

Now let $q \to \infty$ and then, recalling (1.3.3) and also that $\lim_{q\to\infty}(1-\frac{u}{q})^q = e^{-u}$, we have[21]

$$(1.6.2) \qquad \gamma = \int_0^1 \frac{1-e^{-u}}{u} du - \int_1^\infty \frac{e^{-u}}{u} du.$$

The result (1.6.2) probably strikes you as both beautiful and mysterious. But is that all it is? That is, is it just a pretty array of symbols asking only for our admiration, or can we actually *do* something with it? The answer is yes, (1.6.2) is *most* useful. To demonstrate that, consider the evaluation of the simple-looking integral

$$\int_0^\infty e^{-x} \ln(x) dx,$$

a definite integral that, despite looking simple, would give all the usual AP-calculus integration techniques you've seen a real run for their money. (If you don't believe that, try doing it right now.) With (1.6.2), however, you can do this integral as follows.

We begin by splitting the integral into two parts:

$$(1.6.3) \qquad \int_0^\infty e^{-x} \ln(x) dx = \int_0^1 e^{-x} \ln(x) dx + \int_1^\infty e^{-x} \ln(x) dx$$

21. See, for example, Eli Maor, *e: The Story of a Number* (Princeton University Press, 2015).

and then observe that

$$e^{-x} = -\frac{d}{dx}(e^{-x} - 1).$$

(This works because the derivative of a constant (-1) is zero, but why, in particular, -1? Why not some other constant, which would (it would seem) work equally well? Think about this, and I'll ask you again as a challenge problem at the end of this section.) Using this in the first integral on the right in (1.6.3), we have

$$\int_0^1 e^{-x} \ln(x)dx = -\int_0^1 \left\{ \frac{d}{dx}(e^{-x} - 1) \right\} \ln(x)dx.$$

Integrating by parts,[22] this becomes

$$(1.6.4) \quad \int_0^1 e^{-x} \ln(x)dx = -\left[\{(e^{-x} - 1)\ln(x)\}\big|_0^1 - \int_0^1 \frac{e^{-x} - 1}{x}dx \right]$$

$$= -\int_0^1 \frac{1 - e^{-x}}{x}dx.$$

Turning our attention now to the second integral on the right in (1.6.3), and again integrating by parts (with $w = \ln(x)$ and $dz = e^{-x}\,dx$, and so $dw = \frac{1}{x}dx$ and $z = -e^{-x}$), we have

$$(1.6.5) \quad \int_1^\infty e^{-x} \ln(x)dx = \{-e^{-x}\ln(x)\}\big|_1^\infty + \int_1^\infty \frac{e^{-x}}{x}dx = \int_1^\infty \frac{e^{-x}}{x}dx.$$

Putting (1.6.4) and (1.6.5) into (1.6.3) gives us

$$\int_0^\infty e^{-x} \ln(x)dx = -\int_0^1 \frac{1 - e^{-x}}{x}dx + \int_1^\infty \frac{e^{-x}}{x}dx = -\left[\int_0^1 \frac{1 - e^{-u}}{u}du - \int_1^\infty \frac{e^{-u}}{u}du \right]$$

---

22. In the formula $\int w\,dz = (wz) - \int z\,dw$, let $w = \ln(x)$ and $dz = \frac{d}{dx}(e^{-x} - 1)dx$. Then $dw = \frac{1}{x}dx$ and $z = e^{-x} - 1$.

and so, just like that—if we remember (1.6.2)—we see that

$$(1.6.6) \qquad \int_0^\infty e^{-x} \ln(x)dx = -\gamma.$$

We can test our theoretical result of (1.6.6) by doing a direct numerical evaluation of the integral. This is particularly easy to do in *MATLAB*, using that language's *integral* command. The syntax is straightforward, with the arguments of *integral* being simply the integrand, and the upper and lower limits of integration. So, using *MATLAB*'s name for infinity, *inf*, if we type

$$integral(@(x)exp(-x).*log(x),0,inf),$$

then *MATLAB* almost instantly returns the value −0.5772, and (1.6.6) is "confirmed" (or, at least it is to the satisfaction of engineers and physicists).

To end this discussion on integral forms for Euler's constant, let me show you something that you don't usually find in an AP-calculus course, but which is nonetheless within easy reach at that level. Suppose $x \ge 0$, and let $\{x\}$ denote the fractional part of x. For example, $\{7.137\} = 0.137$ and $\{7\} = 0$. Notice that if $\lfloor x \rfloor$ is the *integer* part of x (and so $\lfloor 7.137 \rfloor = 7$), then $x = \lfloor x \rfloor + \{x\}$ and so $\{x\} = x - \lfloor x \rfloor$. Let's now calculate the value of the integral

$$\int_0^1 x^n \left\{\frac{1}{x}\right\} dx, n \ge 0.$$

the existence of which should be obvious, because both x^n and $\{\frac{1}{x}\}$ are each always between 0 and 1 over the entire interval of integration. Don't let the fact that $\frac{1}{x} \to \infty$ as $x \to 0$ distract you. While $\frac{1}{x}$ does indeed blow up as x nears the lower limit, $\{\frac{1}{x}\}$ means *just* the *fractional* part of $\frac{1}{x}$ and that is, for *any* $\frac{1}{x}$, always between 0 and 1.

To start, change variable to $y = \frac{1}{x}$, and so

$$\frac{dy}{dx} = -\frac{1}{x^2}$$

or

$$dx = -x^2 dy = -\frac{dy}{y^2},$$

which says

$$\int_0^1 x^n \left\{ \frac{1}{x} \right\} dx = \int_\infty^1 \frac{1}{y^n} \{y\} \left(-\frac{dy}{y^2} \right) = \int_1^\infty \frac{\{y\}}{y^{n+2}} dy = \sum_{k=1}^\infty \int_k^{k+1} \frac{\{y\}}{y^{n+2}} dy$$

$$= \sum_{k=1}^\infty \int_k^{k+1} \frac{y - \lfloor y \rfloor}{y^{n+2}} dy = \sum_{k=1}^\infty \int_k^{k+1} \frac{y - k}{y^{n+2}} dy,$$

where, in the last step, we use the fact that over the entire open interval k to $k + 1$, the integer part of y is fixed at k. Thus,

(1.6.7) $$\int_0^1 x^n \left\{ \frac{1}{x} \right\} dx = \sum_{k=1}^\infty \int_k^{k+1} \frac{dy}{y^{n+1}} - \sum_{k=1}^\infty k \int_k^{k+1} \frac{dy}{y^{n+2}}.$$

The two integrals on the right of (1.6.7) are each easy to do. In particular, if $n \neq 0$,

$$\int_k^{k+1} \frac{dy}{y^{n+1}} = \frac{1}{-n} y^{-n} \Big|_k^{k+1} = -\frac{1}{ny^n} \Big|_k^{k+1} = -\frac{1}{n} \left[\frac{1}{(k+1)^n} - \frac{1}{k^n} \right]$$

$$= \frac{1}{n} \left[\frac{1}{k^n} - \frac{1}{(k+1)^n} \right]$$

and

$$\int_k^{k+1} \frac{dy}{y^{n+2}} = \frac{1}{-n-1} y^{-n-1} \Big|_k^{k+1} = -\frac{1}{(n+1)} \left[\frac{1}{(k+1)^{n+1}} - \frac{1}{k^{n+1}} \right]$$

$$= \frac{1}{n+1} \left[\frac{1}{k^{n+1}} - \frac{1}{(k+1)^{n+1}} \right].$$

So, as long as $n > 0$ (I'll ask you to treat the $n = 0$ case in Challenge Problem 1.6.2), we have

$$(1.6.8) \qquad \int_0^1 x^n \left\{ \frac{1}{x} \right\} dx = \frac{1}{n} \sum_{k=1}^{\infty} \left[\frac{1}{k^n} - \frac{1}{(k+1)^n} \right]$$

$$- \frac{1}{n+1} \sum_{k=1}^{\infty} k \left[\frac{1}{k^{n+1}} - \frac{1}{(k+1)^{n+1}} \right].$$

The first sum on the right of (1.6.8) is, if we write it out term-by-term,

$$\left[\frac{1}{1^n} - \frac{1}{2^n} \right] + \left[\frac{1}{2^n} - \frac{1}{3^n} \right] + \left[\frac{1}{3^n} - \frac{1}{4^n} \right] + \cdots,$$

which obviously telescopes to $\dfrac{1}{1^n} = 1$, and so (1.6.8) becomes

$$(1.6.9) \qquad \int_0^1 x^n \left\{ \frac{1}{x} \right\} dx = \frac{1}{n} - \frac{1}{n+1} \sum_{k=1}^{\infty} k \left[\frac{1}{k^{n+1}} - \frac{1}{(k+1)^{n+1}} \right].$$

The last sum in (1.6.8)/(1.6.9) is only just a bit more involved. Write

$$\sum_{k=1}^{\infty} k \left[\frac{1}{k^{n+1}} - \frac{1}{(k+1)^{n+1}} \right] = \sum_{k=1}^{\infty} \left[\frac{1}{k^n} - \frac{k}{(k+1)^{n+1}} \right]$$

$$= \sum_{k=1}^{\infty} \left[\frac{1}{k^n} - \frac{1}{(k+1)^n} \left(\frac{k}{k+1} \right) \right]$$

$$= \sum_{k=1}^{\infty} \left[\frac{1}{k^n} - \frac{1}{(k+1)^n} \left(1 - \frac{1}{k+1} \right) \right] = \sum_{k=1}^{\infty} \left[\frac{1}{k^n} - \frac{1}{(k+1)^n} + \frac{1}{(k+1)^{n+1}} \right]$$

and so (1.6.9) becomes

$$(1.6.10) \qquad \int_0^1 x^n \left\{ \frac{1}{x} \right\} dx = \frac{1}{n} - \frac{1}{n+1} \left\{ \sum_{k=1}^{\infty} \left[\frac{1}{k^n} - \frac{1}{(k+1)^n} \right] \right.$$

$$\left. + \sum_{k=1}^{\infty} \frac{1}{(k+1)^{n+1}} \right\}.$$

From before we know that

$$\sum_{k=1}^{\infty}\left[\frac{1}{k^n}-\frac{1}{(k+1)^n}\right]=1,$$

and so (1.6.10) becomes

(1.6.11) $$\int_0^1 x^n\left\{\frac{1}{x}\right\}dx=\frac{1}{n}-\frac{1}{n+1}-\frac{1}{n+1}\sum_{k=1}^{\infty}\frac{1}{(k+1)^{n+1}}.$$

Since

$$\sum_{k=1}^{\infty}\frac{1}{(k+1)^{n+1}}=\frac{1}{2^{n+1}}+\frac{1}{3^{n+1}}+\frac{1}{4^{n+1}}+\cdots=\zeta(n+1)-1,$$

then

$$\int_0^1 x^n\left\{\frac{1}{x}\right\}dx=\frac{1}{n}-\frac{1}{n+1}-\frac{1}{n+1}\left[\zeta(n+1)-1\right]$$

or

(1.6.12) $$\int_0^1 x^n\left\{\frac{1}{x}\right\}dx=\frac{1}{n}-\frac{\zeta(n+1)}{n+1}, n>0.$$

For $n=2$, for example,

$$\int_0^1 x^2\left\{\frac{1}{x}\right\}dx=\frac{1}{2}-\frac{\zeta(3)}{3},$$

which is equal to 0.099314 . . . . If we numerically evaluate the integral in (1.6.12) for $n=2$, we write the integrand as

$$x^2\left(\frac{1}{x}-\left\lfloor\frac{1}{x}\right\rfloor\right)$$

which is coded in *MATLAB* as[23]

$$(x.\wedge 2).*(1./x - floor(1./x)).$$

Thus, a numerical value for the integral can be computed by *MATLAB* by typing

$$integral(@(x)(x.\wedge 2).*(1./x - floor(1./x)),0,1),$$

which returns a value of $0.099314\dots$, in pretty good agreement with (1.6.12).

To see the zeta function showing up in the answer to an integral as exotic as (1.6.12) no doubt may encourage you to think it's that "exotic nature" that is the cause, but in fact, that's not so. As an example to illustrate this, let's do the far less dramatic integral

$$\int_0^1 \frac{\ln^3(1-x^2)}{x}dx,$$

which contains nothing as complicated as a fractional part. To start, make the change of variable $u = x^2$ $(du = 2xdx)$, and so

$$\int_0^1 \frac{\ln^3(1-x^2)}{x}dx = \int_0^1 \frac{\ln^3(1-u)}{x}\frac{du}{2x} = \frac{1}{2}\int_0^1 \frac{\ln^3(1-u)}{u}du.$$

Next, with the second change of variable $v = 1 - u$ $(du = -dv)$, we have

$$\int_0^1 \frac{\ln^3(1-x^2)}{x}dx = \frac{1}{2}\int_1^0 \frac{\ln^3(v)}{1-v}(-dv) = \frac{1}{2}\int_0^1 \frac{\ln^3(v)}{1-v}dv.$$

---

23. The *floor* command *truncates*, that is, rounds downward toward minus infinity, and so returns a value that is the *greatest integer less than or equal to* the argument. The classic *round* command, by contrast, returns the *closest* integer to the argument. For example, *floor*(7.6) = 7, while *round*(7.6) = 8.

Then, since

$$\frac{\ln^3(v)}{1-v} = \ln^3(v)\{1 + v + v^2 + v^3 + \cdots\} = \sum_{n=0}^{\infty} v^n \ln^3(v),$$

we arrive at

$$\int_0^1 \frac{\ln^3(1-x^2)}{x} dx = \frac{1}{2}\int_0^1 \sum_{n=0}^{\infty} v^n \ln^3(v) dv = \frac{1}{2}\sum_{n=0}^{\infty} \int_0^1 v^n \ln^3(v) dv.$$

At this point, we might seem to have merely exchanged our original integral for another integral that is equally challenging. There is, however, an elementary way to do that last integral (if a way not commonly taught as part of AP-calculus), with a result (see the following box)[24] we'll find highly useful in Chapter 4:

(1.6.13) $$\int_0^1 v^n \ln^3(v) dv = \frac{-6}{(n+1)^4}.$$

Thus,

$$\int_0^1 \frac{\ln^3(1-x^2)}{x} dx = \frac{1}{2}\sum_{n=0}^{\infty} \frac{-6}{(n+1)^4} = -3\sum_{k=1}^{\infty} \frac{1}{k^4} = -3\zeta(4) = -\frac{\pi^4}{30},$$

where I've used $\zeta(4) = \frac{\pi^4}{90}$, a result stated without proof in Section 1.2 (we will derive it, using Fourier series, in Section 3.4). *MATLAB* agrees with this result, as

$$-\frac{\pi^4}{30} = -3.2469697\ldots,$$

while *integral(@(x)((log(1-x.^2)).^3)./x,0,1) = −3.2469697 . . . .*

---

24. The idea of evaluating an integral by differentiating it with respect to a parameter (in the integrand and/or in the limits) has become popularly known in recent years as "Feynman's trick," after the mathematical physicist Richard Feynman (whom you'll recall from the discussion of the Feynman-Hibbs integral in the Preface). The trick was well known to mathematicians long before Feynman, however. You can read more on the trick, on Feynman, and on how his name became attached to the trick, in my book, *Inside Interesting Integrals* (Springer, 2020).

We start with the integral $I(n)$, defined as

$$I(n) = \int_0^1 x^n \, dx = \frac{x^{n+1}}{n+1}\Big|_0^1 = \frac{1}{n+1}.$$

In contrast, we can also write

$$I(n) = \int_0^1 e^{\ln(x^n)} \, dx = \int_0^1 e^{n\ln(x)} \, dx.$$

Now, assuming that if we differentiate with respect to the parameter n we can interchange the order of differentiation and integration (not always true, but it is here) then

$$\frac{dI}{dn} = \frac{d}{dn}\int_0^1 e^{n\ln(x)} \, dx = \int_0^1 \frac{d}{dn} e^{n\ln(x)} \, dx = \int_0^1 \ln(x) e^{n\ln(x)} \, dx$$

$$= \int_0^1 \ln(x) e^{\ln(x^n)} \, dx = \int_0^1 x^n \ln(x) \, dx.$$

But since

$$\frac{dI}{dn} = \frac{d}{dn}\left(\frac{1}{n+1}\right) = -\frac{1}{(n+1)^2},$$

we have

$$\int_0^1 x^n \ln(x) \, dx = \int_0^1 v^n \ln(v) \, dv = -\frac{1}{(n+1)^2} = J(n).$$

We now repeat this procedure, that is, we write

$$\frac{dJ}{dn} = \frac{d}{dn}\int_0^1 e^{\ln(v^n)} \ln(v) \, dv = \int_0^1 \frac{d}{dn} e^{n\ln(v)} \ln(v) \, dv$$

$$= \int_0^1 \ln(v) \, e^{n\ln(v)} \ln(v) \, dv = \int_0^1 v^n \ln^2(v) \, dv.$$

But since

$$\frac{dJ}{dn} = \frac{d}{dn}\left\{-\frac{1}{(n+1)^2}\right\} = \frac{2(n+1)}{(n+1)^4} = \frac{2}{(n+1)^3}$$

then

$$\int_0^1 v^n \ln^2(v)\,dv = \frac{2}{(n+1)^3} = K(n).$$

Now, if you repeat this procedure one last time (you should do this), you'll quickly arrive at (1.6.13):

$$\int_0^1 v^n \ln^3(v)\,dv = \frac{-6}{(n+1)^4}.$$

As a challenge, see if you can also derive (1.6.13) via integration by parts.

Challenge Problem 1.6.1: In evaluating the integral of (1.6.3), you'll recall that we wrote $e^{-x} = -\frac{d}{dx}(e^{-x} - 1)$. This is true, but so is $e^{-x} = -\frac{d}{dx}(e^{-x} - 17)$. What's so special about that -1? Hint: Take a close look at the resulting integral in (1.6.4), and ask yourself if that integral would exist for any choice for the constant *other* than -1.

Challenge Problem 1.6.2: If we plug $n = 0$ into the right-hand side of (1.6.12), in an attempt to evaluate $\int_0^1 \{\frac{1}{x}\}dx$, we get the indeterminate result $\infty - \infty$, because $\zeta(1)$ is the divergent harmonic series. The integral, however, clearly *does* exist. So, what is it? A numerical estimate can be computed by *MATLAB* by typing *integral(@(x)(1./x − floor(1./x)),0,1)*, which returns the value 0.4228. From that, can you guess what the exact answer is? (Your guess might be helped by looking instead at $1 − 0.4228 = 0.5772$.) Can you derive the exact answer? Hint: Try plugging $n = 0$ into an earlier step of the derivation of (1.6.12), before the distinction between $n = 0$ and $n > 0$ becomes important.

Challenge Problem 1.6.3: Evaluate the integral $\int_0^1 \frac{\{\ln(1-x)\}^2}{x} dx$, and show that it is equal to $2\zeta(3)$. This integral is encountered in the study of the magnetic moment of the electron and so is of great interest to physicists.[25] Hint: Change variable to $1 - x = e^{-t}$ and then remember Riemann's integral formula in (1.4.24).

1.7 Euler's Constant and the Zeta Function Redux (and the Digamma Function, Too)

In this, the penultimate section of the chapter, I want to take advantage of your new willingness to accept (I hope!) what I earlier called "symbol pushing" (in the discussion on Ramanujan's Master Theorem). We'll actually, in fact, be not quite so outrageous as we were with the RMT. To start, let me write the definition of the gamma function again:

$$(1.7.1) \qquad \Gamma(x) = \int_0^\infty e^{-t} t^{x-1} dt,$$

which is (1.4.1) with a (trivial) change in notation.[26] Next, recalling from the very definition of the exponential that[27]

$$e^{-t} = \lim_{n \to \infty} \left(1 - \frac{t}{n}\right)^n,$$

---

25. Robert Karplus and Norman M. Kroll, "Fourth-Order Corrections in Quantum Electrodynamics and the Magnetic Moment of the Electron," *Physical Review*, September 15, 1949, pp. 846–847.

26. To get (1.7.1) from (1.4.1), simply replace every n in (1.4.1) with an x, and every *original* x in (1.4.1) with a t.

27. See, for example, Maor, *e: The Story of a Number.*

we can write (1.7.1) in a form that might appear, at first glance, to be a step backward (but isn't, as you'll soon see): If we define

(1.7.2) $$\Gamma(x,n) = \int_0^n t^{x-1}\left(1 - \frac{t}{n}\right)^n dt,$$

then

(1.7.3) $$\Gamma(x) = \lim_{n\to\infty} \Gamma(x,n).$$

If we now change variable to $y = \frac{t}{n}$ (and so $dt = n\,dy$), we can write (1.7.2) as

$$\Gamma(x,n) = \int_0^1 (ny)^{x-1}(1-y)^n\, n\,dy$$

or

(1.7.4) $$\Gamma(x,n) = n^x \int_0^1 y^{x-1}(1-y)^n\, dy.$$

If we next integrate by parts,[28] we quickly get

$$\Gamma(x,n) = \frac{1}{x} n^{x+1} \int_0^1 y^x (1-y)^{n-1}\, dy.$$

Since (1.7.4) says that

$$\Gamma(x+1,n-1) = (n-1)^{x+1} \int_0^1 y^x (1-y)^{n-1}\, dy,$$

---

28. Let $u = (1-y)^n$ and $dv = y^{x-1}\,dy$ in $\int_0^1 u\,dv = (uv)\big|_0^1 - \int_0^1 v\,du$.

we have

$$\int_0^1 y^x (1-y)^{n-1} \, dy = \frac{\Gamma(x+1, n-1)}{(n-1)^{x+1}}$$

and so

$$\Gamma(x,n) = \frac{1}{x} n^{x+1} \frac{\Gamma(x+1, n-1)}{(n-1)^{x+1}}$$

or

$$(1.7.5) \qquad \Gamma(x,n) = \frac{1}{x}\left(\frac{n}{n-1}\right)^{x+1} \Gamma(x+1, n-1).$$

To use the recurrence of (1.7.5), we need one more result. Putting $n = 1$ in (1.7.4) yields

$$\Gamma(x,1) = \int_0^1 y^{x-1}(1-y) \, dy = \int_0^1 y^{x-1} \, dy - \int_0^1 y^x \, dy$$

$$= \left(\frac{y^x}{x}\right)\Big|_0^1 - \left(\frac{y^{x+1}}{x+1}\right)\Big|_0^1$$

$$= \frac{1}{x} - \frac{1}{x+1}$$

and so

$$(1.7.6) \qquad \Gamma(x,1) = \frac{1}{x(x+1)}.$$

Now, set $n = 2$ in (1.7.5). This gives

$$\Gamma(x,2) = \frac{1}{x}\left(\frac{2}{1}\right)^{x+1} \Gamma(x+1, 1)$$

where, from (1.7.6),

$$\Gamma(x+1,1)=\frac{1}{(x+1)(x+2)}.$$

Thus,

$$(1.7.7)\qquad \Gamma(x,2)=\frac{1}{x}\left(\frac{2}{1}\right)^{x+1}\frac{1}{(x+1)(x+2)}=\frac{1}{x}2^x\frac{(2)}{(x+1)(x+2)}.$$

Next, set $n=3$ in (1.7.5). This gives

$$\Gamma(x,3)=\frac{1}{x}\left(\frac{3}{2}\right)^{x+1}\Gamma(x+1,2).$$

From (1.7.7) we have

$$\Gamma(x+1,2)=\frac{1}{x+1}\left(\frac{2}{1}\right)^{x+2}\frac{1}{(x+2)(x+3)}$$

and so

$$\Gamma(x,3)=\frac{1}{x}\left(\frac{3}{2}\right)^{x+1}\frac{1}{x+1}\left(\frac{2}{1}\right)^{x+2}\frac{1}{(x+2)(x+3)}$$

or

$$(1.7.8)\qquad \Gamma(x,3)=\frac{1}{x}3^x\frac{(3)(2)}{(x+1)(x+2)(x+3)}.$$

I'll let you repeat this process a few more times if you need more convincing, but the general result is

$$\Gamma(x,n)=n^x\frac{n!}{x(x+1)(x+2)\cdots(x+n)}.$$

That is, using (1.7.3), the gamma function can be written as

$$(1.7.9) \qquad \Gamma(x) = \lim_{n \to \infty} n^x \frac{n!}{x(x+1)(x+2) \cdots (x+n)}$$

which, if you look at it for a while, is truly an amazing statement, one that can be found in a 1729 Euler letter (to the same friend mentioned in Challenge Problem 1.1). As amazing as (1.7.9) is, however, we are not done yet.

Applying some simple algebraic manipulations to (1.7.9) yields

$$\Gamma(x) = \lim_{n \to \infty} n^x \frac{n!}{x(x+1)\left[2\left(\dfrac{x}{2}+1\right)\right]\left[3\left(\dfrac{x}{3}+1\right)\right] \cdots \left[n\left(\dfrac{x}{n}+1\right)\right]}$$

$$= \lim_{n \to \infty} n^x \frac{n!}{x(x+1)\left[\left(\dfrac{x}{2}+1\right)\right]\left[\left(\dfrac{x}{3}+1\right)\right] \cdots \left[\left(\dfrac{x}{n}+1\right)\right]n!}$$

and so

$$(1.7.10) \qquad \Gamma(x) = \lim_{n \to \infty} \frac{n^x}{x \prod_{k=1}^{n}\left(1+\dfrac{x}{k}\right)}.$$

Next, notice that since

$$n+1 = \left(\frac{n+1}{n}\right)\left(\frac{n}{n-1}\right)\left(\frac{n-1}{n-2}\right) \cdots \left(\frac{3}{2}\right)\left(\frac{2}{1}\right)$$

$$= \left(1+\frac{1}{n}\right)\left(1+\frac{1}{n-1}\right)\left(1+\frac{1}{n-2}\right) \cdots \left(1+\frac{1}{2}\right)\left(1+\frac{1}{1}\right),$$

then

$$(n+1)^x = \left(1+\frac{1}{n}\right)^x\left(1+\frac{1}{n-1}\right)^x\left(1+\frac{1}{n-2}\right)^x \cdots \left(1+\frac{1}{2}\right)^x\left(1+\frac{1}{1}\right)^x$$

and so

(1.7.11)
$$(n+1)^x = \prod_{k=1}^{n}\left(1+\frac{1}{k}\right)^x.$$

Finally, since

$$\lim_{n\to\infty}\frac{n^x}{(n+1)^x}=1,$$

we can, with vanishing error, replace n^x in (1.7.10) with $(n+1)^x$ in (1.7.11) to get, in the limit as $n\to\infty$,

(1.7.12)
$$\Gamma(x)=\frac{\prod_{k=1}^{\infty}\left(1+\frac{1}{k}\right)^x}{x\prod_{k=1}^{\infty}\left(1+\frac{x}{k}\right)}.$$

Now, notice that

$$\frac{d}{dx}[\ln\{x\Gamma(x)\}]=\frac{d}{dx}[\ln\{x!\}],$$

where the right-hand side follows from (1.4.4), and since (1.7.12) says

$$x\Gamma(x)=\prod_{k=1}^{\infty}\frac{\left(1+\frac{1}{k}\right)^x}{\left(1+\frac{x}{k}\right)}$$

then

$$\frac{d}{dx}[\ln\{x!\}]=\frac{d}{dx}\ln\left\{\prod_{k=1}^{\infty}\frac{\left(1+\frac{1}{k}\right)^x}{\left(1+\frac{x}{k}\right)}\right\}$$

or

$$\frac{d}{dx}[\ln\{x!\}] = \frac{d}{dx}\left[x\sum_{k=1}^{\infty}\ln\left(1+\frac{1}{k}\right) - \sum_{k=1}^{\infty}\ln\left(1+\frac{x}{k}\right)\right].$$

That is,

(1.7.13) $$\frac{d}{dx}[\ln\{x!\}] = \sum_{k=1}^{\infty}\ln\left(1+\frac{1}{k}\right) - \sum_{k=1}^{\infty}\frac{\frac{1}{k}}{1+\frac{x}{k}}.$$

We can write (1.7.13) as

$$\frac{d}{dx}[\ln\{x!\}] = \lim_{n\to\infty}\sum_{k=1}^{n}\left\{\ln\left(\frac{k+1}{k}\right) - \frac{1}{x+k}\right\}$$

or, since

$$\sum_{k=1}^{n}\ln\left(\frac{k+1}{k}\right) = \ln\left(\frac{2}{1}\right) + \ln\left(\frac{3}{2}\right) + \ln\left(\frac{4}{3}\right) + \cdots + \ln\left(\frac{n+1}{n}\right)$$

$$= \ln\left\{\left(\frac{2}{1}\right)\left(\frac{3}{2}\right)\left(\frac{4}{3}\right)\cdots\left(\frac{n+1}{n}\right)\right\} = \ln(n+1),$$

then

$$\frac{d}{dx}[\ln\{x!\}] = \lim_{n\to\infty}\left\{\ln(n+1) - \sum_{k=1}^{n}\frac{1}{x+k}\right\}$$

$$= \lim_{n\to\infty}\left\{\ln(n+1) - \sum_{k=1}^{n}\frac{1}{k} + \sum_{k=1}^{n}\left(\frac{1}{k} - \frac{1}{x+k}\right)\right\}$$

$$= \lim_{n\to\infty}\left\{\left(\ln(n+1) - \sum_{k=1}^{n}\frac{1}{k}\right) + x\sum_{k=1}^{n}\frac{1}{k(x+k)}\right\}.$$

Recalling the definition of Euler's constant in (1.3.3), we arrive at

$$(1.7.14) \qquad \frac{d}{dx}[\ln\{x!\}] = -\gamma + x\sum_{k=1}^{\infty}\frac{1}{k(x+k)}.$$

We now put (1.7.14) into a form involving the zeta function, as follows. Looking at just the last term on the right, we have

$$x\sum_{k=1}^{\infty}\frac{1}{k(x+k)} = x\sum_{k=1}^{\infty}\frac{1}{k\left(\dfrac{x}{k}+1\right)k} = x\sum_{k=1}^{\infty}\frac{1}{k^2\left(\dfrac{x}{k}+1\right)}$$

$$= x\sum_{k=1}^{\infty}\frac{1}{k^2}\left[1-\left(\frac{x}{k}\right)+\left(\frac{x}{k}\right)^2-\left(\frac{x}{k}\right)^3+\cdots\right], \; -1<x\leq1,$$

and so (1.7.14) becomes

$$(1.7.15) \qquad \frac{d}{dx}[\ln\{x!\}] = -\gamma + \sum_{k=1}^{\infty}\frac{x}{k^2} - \sum_{k=1}^{\infty}\frac{x^2}{k^3}$$

$$+\sum_{k=1}^{\infty}+\frac{x^3}{k^4}-\cdots, \; -1<x\leq1.$$

Thus, integrating indefinitely, with K an arbitrary constant, results in

$$\ln\{x!\} = -\gamma x + \zeta(2)\frac{x^2}{2} - \zeta(3)\frac{x^3}{3} + \zeta(4)\frac{x^4}{4} - \cdots + K.$$

For the case of $x = 0$, we have

$$\ln\{0!\} = \ln\{1\} = 0 = K$$

and so, at last (!),

$$(1.7.16) \qquad \ln\{x!\} = -\gamma x + \sum_{k=2}^{\infty}(-1)^k\frac{\zeta(k)}{k}x^k, \; -1<x\leq1,$$

and we see that (1.7.16) reduces to our earlier result in (1.3.6) for the case of $x = 1$. While (1.3.6) is elegant, it has the flaw of not converging very quickly. With (1.7.16), however, we can have both elegance *and* good convergence (thus allowing for a much-improved estimate of γ).

To see what I mean, let's set $x = \frac{1}{2}$. Then, (1.7.16) says

$$\ln\left\{\left(\frac{1}{2}\right)!\right\} = -\frac{1}{2}\gamma + \sum_{k=2}^{\infty}(-1)^k \frac{\zeta(k)}{k}\left(\frac{1}{2}\right)^k$$

and so, since $(\frac{1}{2})! = \frac{1}{2}\sqrt{\pi}$, we have

$$(1.7.17) \qquad \gamma = 2\left[\sum_{k=2}^{\infty}(-1)^k \frac{\zeta(k)}{k 2^k} - \ln\left(\frac{1}{2}\sqrt{\pi}\right)\right].$$

Because of the 2^k factor in the denominators of the sum's terms, we expect fast convergence. In fact, using just the first 10 terms, we get $\gamma \approx 0.5772$, whereas with (1.3.6), you'll recall the first 10 terms gave the far less accurate 0.5338. The first 100 terms in our fast-converging series give $\gamma = 0.577215664901533$, with all but the last digit correct.

Now, one last development. Returning to (1.7.10), and flipping it upside down,[29]

$$(1.7.18) \qquad \frac{1}{\Gamma(x)} = \lim_{n\to\infty} \frac{x\prod_{k=1}^{n}\left(1+\frac{x}{k}\right)}{n^x}.$$

Writing

$$n^x = e^{\ln(n^x)} = e^{x\ln(n)} = \left\{e^{x\left[\ln(n) - 1 - \frac{1}{2} - \frac{1}{3} - \dots - \frac{1}{n}\right]}\right\} e^{\left[x + \frac{x}{2} + \frac{x}{3} + \dots + \frac{x}{n}\right]}$$

29. *Why flip $\Gamma(x)$?* $\Gamma(x)$ can take on infinite values, because $x\Gamma(x) = x!$ and $|x!| = \infty$ for x a negative integer (recall Challenge Problem 1.4.2). These infinities become zeros of $\frac{1}{\Gamma(x)}$ and so, because $\Gamma(x)$ itself is never zero, $\frac{1}{\Gamma(x)}$ is always finite. This has important implications in advanced applications.

or, remembering (1.3.3), as $n \to \infty$, we can replace the term in curly brackets with e^{-xy}, and so (1.7.18) becomes (as $n \to \infty$)

$$\frac{1}{\Gamma(x)} = \frac{x \prod_{k=1}^{\infty} \left(1 + \dfrac{x}{k}\right)}{e^{-xy} e^{x} e^{x/2} e^{x/3} \cdots}$$

or

(1.7.19)
$$\frac{1}{\Gamma(x)} = xe^{xy} \prod_{k=1}^{\infty} \left(1 + \frac{x}{k}\right) e^{-x/k}.$$

Taking the logarithm of (1.7.19) results in

$$-\ln\{\Gamma(x)\} = \ln(x) + x\gamma + \sum_{k=1}^{\infty} \left\{ \ln\left(1 + \frac{x}{k}\right) - \frac{x}{k} \right\},$$

which, when differentiated, gives what is called the *logarithmic derivative* of $\Gamma(x)$:

$$\frac{\Gamma'(x)}{\Gamma(x)} = -\frac{1}{x} - \gamma - \sum_{k=1}^{\infty} \left\{ \frac{\dfrac{1}{k}}{1 + \dfrac{x}{k}} - \frac{1}{k} \right\}$$

or

(1.7.20) $$\frac{\Gamma'(x)}{\Gamma(x)} = -\frac{1}{x} - \gamma + \sum_{k=1}^{\infty} \left\{ \frac{1}{k} - \frac{1}{k+x} \right\}.$$

The logarithmic derivative of the gamma function is so important in advanced work that mathematicians have given it its own name and symbol: The *digamma function* is defined to be

(1.7.21) $$\psi(x) = \frac{\Gamma'(x)}{\Gamma(x)},$$

and it is interesting to note that, using (1.7.20), it is easy to show $\psi(1) = -\gamma$. That is, the logarithmic derivative of the gamma *function* (i.e., the digamma function) at $x = 1$ is the *number* "minus gamma." Since $\Gamma(1) = 0! = 1$, then $\Gamma'(1) = -\gamma$.

Recalling (1.4.3), we have

$$\Gamma(x + 1) = x\Gamma(x)$$

and so differentiation gives

$$\Gamma'(x + 1) = \Gamma(x) + x\Gamma'(x)$$

or, dividing through by $\Gamma(x)$,

$$(1.7.22) \qquad \frac{\Gamma'(x+1)}{\Gamma(x)} = 1 + x\frac{\Gamma'(x)}{\Gamma(x)}.$$

But since

$$\Gamma(x) = \frac{\Gamma(x+1)}{x}$$

we can write (1.7.22) as

$$\frac{\Gamma'(x+1)}{\dfrac{\Gamma(x+1)}{x}} = 1 + x\frac{\Gamma'(x)}{\Gamma(x)},$$

which reduces to

$$(1.7.23) \qquad \frac{\Gamma'(x+1)}{\Gamma(x+1)} = \frac{1}{x} + \frac{\Gamma'(x)}{\Gamma(x)}.$$

That is, the functional equation of the logarithmic derivative of the gamma function is

$$(1.7.24) \qquad \psi(x+1) = \frac{1}{x} + \psi(x).$$

Starting with $x = n$ (an arbitrary positive integer) in (1.7.24), we can write the following system of n equations:

$$\psi(n+1) = \frac{1}{n} + \psi(n), \psi(n) = \frac{1}{n-1} + \psi(n-1),$$

$$\psi(n-1) = \frac{1}{n-2} + \psi(n-2),\ldots,$$

$$\psi(3) = \frac{1}{2} + \psi(2), \ \psi(2) = \frac{1}{1} + \psi(1) = 1 - \gamma.$$

Successively substituting this chain of expressions, head into tail, we get

$$\psi(n+1) = \frac{1}{n} + \frac{1}{n-1} + \frac{1}{n-2} + \cdots + \frac{1}{2} + 1 - \gamma$$

or

$$(1.7.25) \qquad \psi(n+1) = -\gamma + \sum_{k=1}^{n} \frac{1}{k}, n = 1,2,3,\ldots,$$

which shows how the logarithmic derivative of the gamma function is connected to the harmonic series via Euler's constant.

From Stirling's asymptotic formula for $x!$, as x increases without bound, we have

$$(1.7.26) \qquad x! \sim \sqrt{2\pi} x^{x+\frac{1}{2}} e^{-x},$$

named after the Scottish mathematician James Stirling (1692–1770)—although it is known that the French-born English mathematician Abraham de Moivre (1667–1754) knew an equivalent form at the same time (or even earlier)—who published it in 1730. Factorials get very large very fast (my hand calculator gives up at 70!), and Stirling's formula is quite useful in computing $x!$ for large x. The formula is called *asymptotic* because, while the absolute error in the right-hand side of (1.7.26) in evaluating the left-hand side blows up as $x \to \infty$, the relative error goes to zero as $x \to \infty$ (that's why $\sim$ is used instead of $=$). That is,

$$\lim_{x \to \infty} \frac{x!}{\sqrt{2\pi} x^{x+\frac{1}{2}} e^{-x}} = 1$$

while

$$\lim_{x \to \infty} \left\{ x! - \sqrt{2\pi} x^{x+\frac{1}{2}} e^{-x} \right\} = \infty.$$

For large x, then, since $\Gamma(x+1) = x!$, we can write, with vanishing relative error,

$$\Gamma(x+1) = \sqrt{2\pi} x^{x+\frac{1}{2}} e^{-x}$$

and so, taking logarithms,

$$\ln\{\Gamma(x+1)\} = \ln(\sqrt{2\pi}) + \left(x + \frac{1}{2}\right) \ln(x) - x,$$

and then differentiating, we have

$$\frac{d}{dx} \ln\{\Gamma(x+1)\} = \psi(x+1) = \left(x + \frac{1}{2}\right) \frac{1}{x} + \ln(x) - 1$$

$$= 1 + \frac{1}{2x} + \ln(x) - 1.$$

That is,

$$\lim_{x\to\infty}\psi(x+1)=\lim_{x\to\infty}\left[\frac{1}{2x}+\ln(x)\right]$$

or, writing n instead of x,

(1.7.27) $\qquad \lim_{n\to\infty}\psi(n+1)=\lim_{n\to\infty}\left[\frac{1}{2n}+\ln(n)\right].$

Combining (1.7.25) and (1.7.27),

$$\lim_{n\to\infty}\left[\frac{1}{2n}+\ln(n)\right]=-\gamma+\lim_{n\to\infty}\sum_{k=1}^{n}\frac{1}{k}$$

or

$$\gamma+\lim_{n\to\infty}\frac{1}{2n}=\lim_{n\to\infty}\left[\sum_{k=1}^{n}\frac{1}{k}-\ln(n)\right]$$

or, as $\lim_{n\to\infty}\frac{1}{2n}=0$, we have

$$\lim_{n\to\infty}\left[\sum_{k=1}^{n}\frac{1}{k}-\ln(n)\right]=\gamma,$$

which is, of course, (1.3.3).

In addition to being important in physics (see notes 13 and 25 again), the zeta function makes a similarly remarkable appearance in computer science. The underlying theory of modern digital systems implementing data encryption relies on the extreme difficulty in factoring very large numbers, where "very large" means numbers that have several hundred digits to them. A technical problem that soon arises in the theory of data encryption is that of determining whether a collection of positive integers is relatively prime. If each integer in the collection is factored into a product of primes (see note 3 again), then the integers are said to be relatively prime if there is no prime factor common to all the products. For example, $8 = (2)(2)(2)$, $9 = (3)(3)$, and $12 = (2)(2)(3)$ are relatively prime. Another way of expressing this is that the numbers are relatively prime if their greatest common divisor is 1. Now, suppose we select, at random, k integers from all the integers from 1 to n. It can then be shown that the probability that the k integers are relatively prime, in the limit as $n \to \infty$, is equal to $1/\zeta(k)$. For $k = 2$ and $k = 3$, these probabilities are 0.6079 and 0.8319, respectively, and these values are easily confirmed with computer simulations. You can find a discussion of how to do that in my book, *How to Fall Slower Than Gravity* (Princeton University Press, 2018), pp. 137–146. This provides yet another illustration of the intimate connection between the zeta function and the primes.

Challenge Problem 1.7.1: The reason I set $x = \frac{1}{2}$ in (1.7.16) to get (1.7.17) is because we have an exact expression for $(\frac{1}{2})$! Setting x even smaller would improve the speed of convergence even more but, unfortunately, there are no other known exact expressions for x! for $|x| \leq 1$. There are, however, *integrals* we can evaluate for a given $|x| \leq 1$ to get x! For example, your problem here is to show that $(\frac{1}{n})! = \int_0^\infty e^{-u^n} du, n \geq 0$.

Challenge Problem 1.7.2: Confirm the claim that $\psi(1) = \Gamma'(1) = -\gamma$ (the value of the digamma function at $x = 1$ is negative gamma). Hint: Notice that the terms of the sum of (1.7.20) telescope when $x = 1$.

Challenge Problem 1.7.3: You'll recall that in (1.6.6), we showed $\int_0^\infty e^{-x} \ln(x) dx = -\gamma$. The next step up in complexity is the integral $\int_0^\infty e^{-x} \{\ln(x)\}^2 dx$. You can find this definite integral in any good math table, but see if you can derive $\int_0^\infty e^{-x} \{\ln(x)\}^2 dx = \gamma^2 + \zeta(2)$. Hint:

Start with $I(m) = \int_0^\infty x^m e^{-x} dx$, and think about how to express $I(m)$ as the second derivative of the gamma function. Then, remember the digamma function.

Challenge Problem 1.7.4: Writing $\Gamma(x + 1) = x!$, we have from (1.7.16) that $\ln\{\Gamma(x+1)\} = -\gamma x + \Sigma_{k=2}^{\infty}(-1)^k \frac{\zeta(k)}{k} x^k$. Exponentiating the left-hand side gives us $e^{\ln\{\Gamma(x+1)\}} = \Gamma(x + 1)$. Exponentiate the right-hand side of the equation, and then use the power series expansions of the resulting exponentials to find the first five terms of the power series expansion for $\Gamma(x + 1)$ about $x = 0$ (the Taylor series). Hint: Confirm that the first three terms are $\Gamma(x+1) = 1 - \gamma x + \frac{1}{2}(\gamma^2 + \frac{\pi^2}{6})x^2$, and then find the coefficients of the next two terms of the Taylor series, that is, the coefficient of x^3 and the coefficient of x^4.

Challenge Problem 1.7.5: Returning to Euler's identity (see Challenge Problem 1.1.1 again), show that i^i is real, a calculation first done by Euler in 1746. That is, an imaginary number ($i = \sqrt{-1}$) raised to an imaginary power, despite sounding pretty complicated, can nonetheless be a *real* number. Hint: Write $i = e^{i\frac{\pi}{2}}$ (because $e^{i\frac{\pi}{2}} = \cos(\frac{\pi}{2}) + i\sin(\frac{\pi}{2}) = 0 + i(1) = i$).

1.8 Calculating $\zeta(3)$

Right after stating the defining sum for $\zeta(3)$ in (1.2.5), I made the casual comment that the numerical value of zeta-3 "is easily calculated," and then I zipped off to a discussion of the zeta function. Well, as you perhaps have been wondering ever since (1.2.5), is the calculation of $\zeta(3)$ as simple as I made it out to be, that is, as being nothing but the evaluation of the terms in (1.2.5)? For a mathematician doing purely theoretical work, that probably is the case, but for engineers and physicists who need an actual number, the defining expression for $\zeta(3)$ is really not so helpful. In fact, using the first 10 terms of (1.2.5) gives a value of $1.19753\ldots$, of which there is *not even a single digit* to the right of the decimal point that is correct! What is needed is an expression that converges to $\zeta(3)$ faster than the $1/n^3$ rate of (1.2.5). So, let me close this first chapter by showing you a way to achieve that, using a clever idea due to the German mathematician Ernst Kummer (1810–1893).[30]

We start by establishing the claim

$$\alpha = \sum\nolimits_{n=2}^{\infty} \frac{1}{(n-1)n(n+1)} = \frac{1}{4}. \tag{1.8.1}$$

We do this by making the partial fraction expansion

$$\frac{1}{(n-1)n(n+1)} = \frac{A}{n-1} + \frac{B}{n} + \frac{C}{n+1}, \tag{1.8.2}$$

where A, B, and C are constants. We can determine the value of A by multiplying through (1.8.2) by $n - 1$ and then setting $n = 1$. That is,

$$\frac{1}{n(n+1)}\Big|_{n=1} = A + B\frac{n-1}{n}\Big|_{n=1} + C\frac{n-1}{n+1}\Big|_{n=1} = A = \frac{1}{2}.$$

Similarly, for B we multiply through (1.8.2) by n and then set $n = 0$ to get

$$\frac{1}{(n-1)(n+1)}\Big|_{n=0} = A\frac{n}{n-1}\Big|_{n=0} + B + C\frac{n}{n+1}\Big|_{n=0} = B = -1.$$

And finally, for C we multiply through (1.8.2) by $n + 1$ and then set $n = -1$ to get

$$\frac{1}{(n-1)n}\Big|_{n=-1} = A\frac{n+1}{n-1}\Big|_{n=-1} + B\frac{n+1}{n}\Big|_{n=-1} + C = C = \frac{1}{2}.$$

30. We have, of course, Markov's series for $\zeta(3)$ that I gave you in the box at the end of Section 1.2. And it's a very good series, too, as using just the first 10 terms gives the estimate 1.202056900 . . ., which has the first eight decimal digits correct. But that series is simply stated; what I want to do here is show you how to *derive* a simple yet fast-converging series for $\zeta(3)$.

Thus,

$$\frac{1}{(n-1)n(n+1)} = \frac{\frac{1}{2}}{n-1} - \frac{1}{n} + \frac{\frac{1}{2}}{n+1} = \frac{1}{2}\left(\frac{1}{n-1} - \frac{2}{n} + \frac{1}{n+1}\right)$$

and so (1.8.1) becomes

$$\alpha = \frac{1}{2}\sum_{n=2}^{\infty}\left(\frac{1}{n-1} - \frac{2}{n} + \frac{1}{n+1}\right)$$

or

(1.8.3) $$\alpha = \frac{1}{2}\left[\left(1 - 1 + \frac{1}{3}\right) + \left(\frac{1}{2} - \frac{2}{3} + \frac{1}{4}\right) + \left(\frac{1}{3} - \frac{2}{4} + \frac{1}{5}\right) + \left(\frac{1}{4} - \frac{2}{5} + \frac{1}{6}\right)\right.$$
$$\left. + \left(\frac{1}{5} - \frac{2}{6} + \frac{1}{7}\right) + \cdots\right].$$

If you carefully examine the terms in the square brackets on the right of (1.8.3), you'll see that they all self-cancel (the series telescopes) except for the $\frac{1}{2}$ in the second pair of parentheses. So $\alpha = \frac{1}{4}$, and (1.8.1) is established.

Now, continuing, let c be an arbitrary constant (arbitrary for now, but we'll soon give it a quite specific value). We can then write

$$\sum_{n=1}^{\infty}\frac{1}{n^3} + c\alpha = 1 + \sum_{n=2}^{\infty}\frac{1}{n^3} + \frac{c}{4}$$

$$= 1 + \sum_{n=2}^{\infty}\frac{1}{n^3} + \sum_{n=2}^{\infty}\frac{c}{(n-1)n(n+1)}$$

$$= 1 + \sum_{n=2}^{\infty}\left\{\frac{1}{n^3} + \frac{c}{(n-1)n(n+1)}\right\} = 1 + \sum_{n=2}^{\infty}\frac{(n-1)n(n+1) + cn^3}{n^4(n-1)(n+1)}$$

$$= 1 + \sum_{n=2}^{\infty}\frac{(n-1)(n+1) + cn^2}{n^3(n-1)(n+1)}.$$

That is,

$$(1.8.4) \qquad \sum_{n=1}^{\infty}\frac{1}{n^3}+c\frac{1}{4}=1+\sum_{n=2}^{\infty}\frac{n^2-1+cn^2}{n^3(n-1)(n+1)}.$$

Now, in (1.8.4) set the up-to-now arbitrary constant c equal to -1. Then,

$$\sum_{n=1}^{\infty}\frac{1}{n^3}-\frac{1}{4}=\zeta(3)-\frac{1}{4}=1+\sum_{n=2}^{\infty}\frac{-1}{n^3(n-1)(n+1)}$$

or finally,

$$(1.8.5) \qquad \zeta(3)=\frac{5}{4}-\sum_{n=2}^{\infty}\frac{1}{n^3(n^2-1)},$$

which converges as n^{-5}, which is much faster than is the n^{-3} convergence rate of (1.2.5).

As an example, using the first 10 terms of (1.8.5) gives $1.20207\ldots$, which has the first four decimal digits correct, compared (you'll recall) to *no* correct decimal digits for (1.2.5). If we extend our calculations to include the first 100 terms, then (1.2.5) gives a value of $1.20200\ldots$, which now has the first four decimal digits correct, but including the first 100 terms of (1.8.5) gives $1.202056905\ldots$, which has the first eight decimal digits correct. Using the same number of terms, the improvement in the rate of convergence of (1.8.5) over that of (1.2.5) is dramatic.

Challenge Problem 1.8.1: In a calculation first done in 1879 by the French mathematician Eugène Lionnet (1805–1884), it was shown that $1+2\sum_{n=1}^{\infty}\frac{1}{(4n)^3-4n}=\frac{3}{2}\ln(2)$. Can you see how to do this? Hint: Start by making the partial fraction expansion $\frac{1}{(4n)^3-4n}=\frac{1}{4n\{(4n)^2-1\}}$ $=\frac{1}{4n(4n-1)(4n+1)}=\frac{A}{4n}+\frac{B}{4n-1}+\frac{C}{4n+1}$, where A, B, and C are particular constants (determine them, using the approach in the text for calculating the partial fraction expansion that speeded up the calculation of $\zeta(3)$). Then use (1.3.5). To give you confidence in this claim,

74

Chapter 1

the left-hand side evaluates (using the first 1,000 terms of the sum) as $1.03972075\ldots$, which agrees pretty well with $\frac{3}{2}\ln(2) = 1.03972077\ldots$.

Challenge Problem 1.8.2: Calculate the value of $\sum_{n=1}^{\infty} \frac{1}{n(n+1)(n+2)(n+3)}$. Hint: Make a partial fraction expansion.

Challenge Problem 1.8.3: As a continuation of the last problem, see if you can make some headway with the general question: What is the value of $\sum_{n=1}^{\infty} \frac{1}{n(n+1)(n+2)\cdots(n+p)}$ for p any given positive integer? The partial fraction approach becomes increasingly messy as the number of factors in the denominator increases, and you might need to go down a different path. This problem is sufficiently challenging that I think it deserves more than a brief discussion in the Solutions section, and so, instead, I'll show you one way to attack this problem at the end of Chapter 2, after we have developed some additional technical results there. This is an interesting problem, because, with the result, additional series having increasingly faster rates of convergence can be found for the calculation of $\zeta(3)$.

More Wizard Math and the Zeta Function ζ(s)

2.1 Euler's Infinite Series for ζ(2)

What's your reaction to the claims that

(2.1.1)
$$1-1+1-1+1-1+\cdots=\frac{1}{2}$$

(2.1.2)
$$1-2+3-4+5-6+\cdots=\frac{1}{4}$$

(2.1.3)
$$1-2^2+3^2-4^2+5^2-6^2+\cdots=0$$

(2.1.4)
$$1-2^3+3^3-4^3+5^3-6^3+\cdots=-\frac{1}{8}$$

and that other similar, seemingly divergent infinite sums (involving ever-increasing powers) all have, against all "common sense," definite *finite* values? Do you think that only a slightly unbalanced (okay, maybe a bit more than just slightly unbalanced) mind would make such claims? If so, you'd be wrong, as these series (and others, even stranger) appeared in papers authored by Euler (for an example of an even weirder Euler series than the ones I've listed, see Challenge Problem 2.1.1). Here's how he reasoned for the above four series.

Starting with the polynomial

$$(2.1.5) \qquad P_0(x) = 1 + x + x^2 + x^3 + x^4 + x^5 + x^6 + \cdots = \frac{1}{1-x}$$

(the last equality is easily confirmed by either cross-multiplying or by directly doing the long division), Euler set $x = -1$ in (2.1.5) to immediately arrive at (2.1.1):

$$1 - 1 + 1 - 1 + 1 - 1 + \cdots = \frac{1}{2},$$

which does have (sort of) a certain plausibility to it. After all, as you proceed from left to right on the left-hand side, computing the partial sums as you go, you see those sums flip back and forth between 1 and 0, and $\frac{1}{2}$ is the *average* value. This series was, in fact, studied decades before Euler by the Italian mathematician Guido Grandi (1671–1742), who discussed it in a 1703 book.

Euler next calculated

$$(2.1.6) \qquad P_1(x) = x \frac{d}{dx} P_0(x) = x \frac{d}{dx} \left(\frac{1}{1-x} \right).$$

That is, he wrote

$$(2.1.7) \qquad x + 2x^2 + 3x^3 + 4x^4 + 5x^5 + 6x^6 + \cdots = \frac{x}{(1-x)^2}$$

and, by setting $x = -1$, he arrived at

$$1-2+3-4+5-6+\cdots=\frac{1}{4}.$$

This establishes the claim in (2.1.2). (Notice that setting $x = 1$ gives $1 + 2 + 3 + 4 + 5 + \ldots = \infty$, a claim that everybody would certainly agree is true and that Euler found to be convincing evidence that the procedure is consistently okay.[1]) Then he repeated the process, over and over. That is, he next wrote

$$(2.1.8) \qquad P_2(x) = x\frac{d}{dx}P_1(x) = x\frac{d}{dx}\left[\frac{x}{(1-x)^2}\right]$$

and so on, from which the other claims quickly follow (with $x = -1$). I'm not going to continue with this thread of Euler's analysis, as we'll soon develop his results in a less audacious way.[2]

If these series seem "odd" to you, just imagine what must have been the reaction of a world-famous mathematician who, upon opening a letter from a stranger, read the claim

$$1+2+3+4+\cdots=-\frac{1}{12}.$$

---

1. But suppose, you may already be wondering, that Euler had instead set $x = 2$ in (2.1.5)? (Remember, all concerns over convergence are out the window!) Then he would have arrived at the claim $1 + 2 + 4 + 8 + \ldots = -1$ which, term-by-term, is at least as large as the sum resulting from setting $x = 1$. That is, $-1 > \infty$, which may be a bit of a surprise. Nonetheless, that's just what Euler concluded! (Euler could, indeed, be bold—to say the least.) You can read more on Euler's reasoning and how he justified what he did, in Euler's own words, in an English translation by E. J. Barbeau and P. J. Leah of his 1754/55 paper "On Divergent Series": see Barbeau and Leah, "Euler's 1760 Paper on Divergent Series," *Historia Mathematica*, May 1976, pp. 141–160 (the difference in dates is the difference between when Euler presented his findings and when they were published). See also Morris Kline, "Euler and Infinite Series," *Mathematics Magazine*, November 1983, pp. 307–314.

2. If you are interested in the details of how Euler proceeded with these particular calculations, you can find a nice discussion (with an occasional typographical error in the numbers, so be alert) in Raymond Ayoub, "Euler and the Zeta Function," *American Mathematical Monthly*, December 1974, pp. 1067–1086.

You'll recall that Euler had declared, from (2.1.7), that the sum on the left is infinity, which makes sense. But $-\frac{1}{12}$? The sum of all the *positive integers* is a *negative fraction*? That's simply crazy, right? And most people would have instantly chucked that letter into the trash, dismissing the writer (with either a curse or a sneer, or perhaps both) as a pathetic lunatic. But both the writer and the mathematician were unusual men, and so that was *not* the fate of the letter. The author was the soon-to-be-discovered Indian genius Ramanujan (recall Section 1.5), and the mathematician was G. H. Hardy in England. Writing to Hardy in January 1913, Ramanujan (then an obscure clerk in Madras) had sent what he thought to be some of his best discoveries, in an attempt to obtain Hardy's support. It is a testament to Hardy's genius that he quickly made sense of Ramanujan's series.

What the clerk meant (but had expressed badly) is understood by writing the sum as

$$1+2+3+4+\cdots = \frac{1}{\dfrac{1}{1}} + \frac{1}{\dfrac{1}{2}} + \frac{1}{\dfrac{1}{3}} + \frac{1}{\dfrac{1}{4}} + \cdots = \frac{1}{1^{-1}} + \frac{1}{2^{-1}} + \frac{1}{3^{-1}} + \frac{1}{4^{-1}} + \cdots,$$

which is, formally, $\zeta(-1)$. As Hardy later discovered, at the time he wrote, Ramanujan had never even heard of the zeta function, but in some incredible way had nevertheless calculated the value of the zeta function $\zeta(s)$—as defined in (1.2.6)—for $s = -1$. Of course, (1.2.6) doesn't even converge for $s \leq 1$, and so there is an immediate question about what $\zeta(-1)$ could mean,[3] but we've been pretty fast and loose with divergent series so far, so let's plunge boldly onward and see if we can make sense of $\zeta(-1)$.

In fact, Euler also could have calculated $\zeta(-1)$ before his death in 1783 as follows. We start with what is called the *alternating zeta function*:

---

3. To put your mind at ease, in a more advanced treatment than this book offers, the introduction of a complex-valued s, and the technique of *analytic continuation*, does indeed bring $\zeta(-1)$ into the world of acceptable mathematics.

$$(2.1.9) \qquad \eta(s) = \sum_{k=1}^{\infty} \frac{(-1)^{k+1}}{k^s} = \frac{1}{1^s} - \frac{1}{2^s} + \frac{1}{3^s} - \frac{1}{4^s} + \cdots.$$

(This is also called *Dirichlet's eta function*.) The eta function is, obviously, quite similar to the zeta function, and indeed, there is a simple relationship between $\eta(s)$ and $\zeta(s)$. Since

$$\zeta(s) = \frac{1}{1^s} + \frac{1}{2^s} + \frac{1}{3^s} + \frac{1}{4^s} + \cdots$$

then

$$\frac{2}{2^s}\zeta(s) = \frac{2}{2^s} + \frac{2}{4^s} + \frac{2}{6^s} + \cdots$$

and so, by inspection, we have

$$(2.1.10) \qquad \eta(s) = \zeta(s) - \frac{2}{2^s}\zeta(s) = (1 - 2^{1-s})\zeta(s).$$

In particular, for $s = -1$,

$$\eta(-1) = (1 - 2^2)\zeta(-1) = -3\zeta(-1)$$

or

$$\zeta(-1) = -\frac{1}{3}\eta(-1) = -\frac{1}{3}\left[\frac{1}{1^{-1}} - \frac{1}{2^{-1}} + \frac{1}{3^{-1}} - \frac{1}{4^{-1}} + \cdots\right]$$

or

$$\zeta(-1) = -\frac{1}{3}\left[1 - 2 + 3 - 4 + \cdots\right].$$

But the divergent series in the brackets is (2.1.2) and, as Euler showed (and so did we, earlier in this section), is "equal" to $\frac{1}{4}$, and so, just like that, we have Ramanujan's result:

$$(2.1.11) \qquad\qquad \zeta(-1) = -\frac{1}{12}.$$

Now this is, without a doubt, pretty fast-and-loose math! Keep reading, however, and you'll see that we'll derive $\zeta(-1)$ in a far more satisfactory way later in the book (and we'll get the same result).

The reason I show you all this here is simply to illustrate the no-tricks-too-outlandish fashion in which Euler (and other similarly unrestrained mathematicians of his day) worked. In the rest of this section, you'll see, in particular, how Euler's devilishly clever mind derived the fast-convergence series for $\zeta(2)$ that I mentioned in the second box of Section 1.2, where I promised you an eventual derivation (and so here it is). Euler's analysis that follows appeared in 1731, three years before he finally solved for the *exact* value of $\zeta(2)$.

Euler started with the elementary observation

$$\int \frac{dx}{1-x} = -\ln(1-x) + K,$$

where K is the arbitrary constant of indefinite integration. So,

$$-\ln(1-x) + K = \int (1 + x + x^2 + x^3 + \cdots)dx = x + \frac{1}{2}x^2 + \frac{1}{3}x^3 + \frac{1}{4}x^4 + \cdots$$

or, since at $x = 0$ we have $-\ln(1) + K = 0 + K = 0$, which means $K = 0$, then

$$(2.1.12) \qquad \ln(1-x) = -x - \frac{1}{2}x^2 - \frac{1}{3}x^3 - \frac{1}{4}x^4 - \cdots.$$

From (2.1.12) it immediately follows that

$$\frac{\ln(1-x)}{x} = -\left[1 + \frac{1}{2}x + \frac{1}{3}x^2 + \frac{1}{4}x^3 + \cdots\right] = -\sum_{n=1}^{\infty}\frac{x^{n-1}}{n}.$$

Thus,

$$\int_0^1 \frac{\ln(1-x)}{x}dx = -\int_0^1 \sum_{n=1}^{\infty}\frac{x^{n-1}}{n}dx = -\sum_{n=1}^{\infty}\frac{1}{n}\int_0^1 x^{n-1}dx$$

$$= -\sum_{n=1}^{\infty}\frac{1}{n}\left(\frac{x^n}{n}\right)\Big|_0^1 = -\sum_{n=1}^{\infty}\frac{1}{n^2}$$

or

(2.1.13) $$\int_0^1 \frac{\ln(1-x)}{x}dx = -\zeta(2).$$

Now, change variable to $t = 1 - x$ (and so $dx = -dt$). Then

$$-\zeta(2) = \int_1^0 \frac{\ln(t)}{1-t}(-dt) = \int_0^1 \frac{\ln(t)}{1-t}dt$$

or, splitting the last integral into two parts (where x is an arbitrary value between 0 and 1),

(2.1.14) $$-\zeta(2) = \int_0^x \frac{\ln(t)}{1-t}dt + \int_x^1 \frac{\ln(t)}{1-t}dt = I_1 + I_2.$$

The trick to arriving at Euler's fast-converging series is to do the two integrals, I_1 and I_2, in different ways (even though they are actually the same, with only their limits being different). Let's do I_2 first, by changing variable to $u = 1 - t$ (and so $dt = -du$). Thus,

$$I_2 = \int_x^1 \frac{\ln(t)}{1-t}dt = \int_{1-x}^0 \frac{\ln(1-u)}{u}(-du) = \int_0^{1-x}\frac{\ln(1-u)}{u}du$$

or, remembering our earlier result for $\frac{\ln(1-x)}{x}$ immediately following (2.1.12),

$$I_2 = -\int_0^{1-x} \left[1 + \frac{1}{2}u + \frac{1}{3}u^2 + \frac{1}{4}u^3 + \cdots\right] du$$

$$= -\left[u + \frac{u^2}{2^2} + \frac{u^3}{3^3} + \frac{u^4}{4^4} + \cdots\right]\Big|_0^{1-x}$$

or

(2.1.15) $$I_2 = -\sum_{n=1}^{\infty} \frac{(1-x)^n}{n^2}.$$

For I_1, Euler expanded the integrand in a power series to write

$$I_1 = \int_0^x \frac{\ln(t)}{1-t} dt = \int_0^x \ln(t)[1 + t + t^2 + t^3 + \cdots] dt,$$

which he then evaluated using integration by parts. That is, letting $u = \ln(t)$ and $dv = [1 + t + t^2 + t^3 + \cdots]\, dt$

in

$$\int_0^x u\, dv = (uv)\Big|_0^x - \int_0^x v\, du,$$

he computed

$$\frac{du}{dt} = \frac{1}{t} \left(\text{and so } du = \frac{dt}{t}\right)$$

and

$$v = t + \frac{1}{2}t^2 + \frac{1}{3}t^3 + \frac{1}{4}t^4 + \cdots,$$

from which it follows that

$$I_1 = \int_0^x \frac{\ln(t)}{1-t}\,dt = \left[\ln(t)\left\{t + \frac{1}{2}t^2 + \frac{1}{3}t^3 + \frac{1}{4}t^4 + \cdots\right\}\right]\Big|_0^x$$

$$- \int_0^x \left(1 + \frac{1}{2}t + \frac{1}{3}t^2 + \frac{1}{4}t^3 + \cdots\right)dt$$

$$= \ln(x)\left[x + \frac{1}{2}x^2 + \frac{1}{3}x^3 + \frac{1}{4}x^4 + \cdots\right] - \left[t + \frac{t^2}{2^2} + \frac{t^3}{3^2} + \frac{t^4}{4^2} + \cdots\right]\Big|_0^x$$

$$= \ln(x)\left[x + \frac{1}{2}x^2 + \frac{1}{3}x^3 + \frac{1}{4}x^4 + \cdots\right] - \sum_{n=1}^{\infty} \frac{x^n}{n^2}.$$

From (2.1.12) we have

$$\left[x + \frac{1}{2}x^2 + \frac{1}{3}x^3 + \frac{1}{4}x^4 + \cdots\right] = -\ln(1-x)$$

and so

$$(2.1.16) \qquad I_1 = \int_0^x \frac{\ln(t)}{1-t}\,dt = -\ln(x)\ln(1-x) - \sum_{n=1}^{\infty} \frac{x^n}{n^2}.$$

Combining (2.1.14), (2.1.15), and (2.1.16), we have

$$-\zeta(2) = -\ln(x)\ln(1-x) - \sum_{n=1}^{\infty} \frac{x^n}{n^2} - \sum_{n=1}^{\infty} \frac{(1-x)^n}{n^2}.$$

Euler's final step was to set $x = \frac{1}{2}$, which gives

$$\zeta(2) = \ln\left(\frac{1}{2}\right)\ln\left(\frac{1}{2}\right) + \sum_{n=1}^{\infty} \frac{1}{2^n n^2} + \sum_{n=1}^{\infty} \frac{1}{2^n n^2}$$

or, finally,

(2.1.17) $\zeta(2) = \{\ln(2)\}^2 + \sum_{n=1}^{\infty} \dfrac{1}{2^{n-1}n^2}$,

the series I gave you in Section 1.2 that converges much more rapidly than does the original defining series for $\zeta(2)$. Indeed, from (2.1.12), Euler knew that

$$\ln(2) = -\ln\left(1 - \frac{1}{2}\right) = \frac{1}{2} + \frac{1}{2}\left(\frac{1}{2}\right)^2 + \frac{1}{3}\left(\frac{1}{2}\right)^3 + \cdots = \sum_{n=1}^{\infty} \frac{1}{n2^n},$$

which, using just the first 20 terms, gives

$$\{\ln(2)\}^2 = 0.480453\ldots,$$

and so the first 20 terms of the sum in (2.1.17) generate a value of 1.164481 . . . , giving an estimate for zeta-2 of $\zeta(2) = 1.644934\ldots$, an estimate that compares very nicely with the exact value of $\frac{\pi^2}{6} = 1.644934\ldots$. Using the first 20 terms of the original defining series for $\zeta(2)$, in contrast, gives 1.596163 . . . , which is not nearly as good an estimate.

Challenge Problem 2.1.1: See if you can "make sense" of the claim (made in the 1760 paper by Euler on divergent series mentioned in note 1) that the divergent series

$$S = \sum_{k=0}^{\infty} (-1)^k k! = 1 - 1 + 2 - 6 + 24 - 120 + 720 - \cdots$$

has a finite value. Hint: Since (1.4.1) says $\Gamma(n) = \int_0^{\infty} e^{-x}x^{n-1}dx$, and since $\Gamma(n) = (n-1)!$ by (1.4.4), then start by writing $k! = \int_0^{\infty} x^k e^{-x}dx$, and put this integral into the given summation for S. Estimate the numerical value of S to at least three decimal places.

Challenge Problem 2.1.2: In 1826 Abel (see note 18 in Chapter 1) proved a theorem that the French mathematician Siméon-Denis Poisson (1781–1840) then used to define a summation method for the infinite series $S = u_0 + u_1 + u_2 + u_3 + \ldots$, even if S is divergent: find a series $T = u_0 + u_1 x + u_2 x^2 + u_3 x^3 + \ldots$, sum it, calculate $\lim_{x \to 1}$

T, and *define* the result (called, despite Poisson's role, the *Abelian sum*) to be the value of S. Can you use this idea to find the Abelian sum value of $S = 1 - 2 + 4 - 8 + 16 - 32 + \cdots = ?$ (You may be amused to learn that Abel was always suspicious of divergent series, and he is famous in mathematics for the observation that "Divergent series are in general the work of the devil, and it is shameful to base any demonstration whatever on them." The modern view of divergent series is a bit less harsh.)

Challenge Problem 2.1.3: Calculate the values of $\zeta(0)$, $\zeta(-2)$, $\zeta(-3)$, and $\zeta(1/2)$. Hint: For $\zeta(0)$, use the eta function and (2.1.1). For $\zeta(\frac{1}{2})$, think of using a computer.

2.2 The Beta Function and the Duplication Formula

We start by defining the *beta function*:

$$(2.2.1) \qquad B(m,n) = \int_0^1 x^{m-1}(1-x)^{n-1}\, dx, \quad m>0, n>0,$$

which is intimately related to the gamma function, as I'll now show you. The pure mathematical definition of the gamma function was introduced in the previous chapter, but perhaps more on the historical origin for the integral of (1.4.1) is now appropriate. Euler's interest in the gamma function originated in the question of how to extend the factorial function from the case of just positive integer arguments to the more general case of *any* real argument. This is an example of what mathematicians call an *interpolation* problem, a type of problem that has occurred numerous times in mathematics. Two examples of it are found in the algebra of exponents and in fractional differentiation. The first case is relatively easy (in hindsight). If n is a positive integer, then a^n is the product of a's and so, for example, $a^3 = (a)(a)(a)$. It was understood that $a^0 = 1$. But what could $a^{3\frac{1}{2}}$ mean? How do you multiply three *and a half a*'s together? Well, we of course know today how exponents work, and so $a^{3\frac{1}{2}} = a^{3+\frac{1}{2}} = a^3 a^{\frac{1}{2}} = a^3\sqrt{a}$. Yes, this all looks obvious now, but it took the genius of Newton to make that so.

The fractional differentiation interpolation problem is just a bit more complicated. The German mathematician Gottfried Wilhelm von Leibnitz (1646–1716) wrote D^n to denote the operation of differentiating n successive times (see note 15 in Chapter 1). It is understood that $D^0 = 1$ means *don't* differentiate, and D^{-n} denotes the inverse operation of *integrating n* successive times. That all seems pretty straightforward, assuming n is a non-negative integer, but what if $n = \frac{1}{2}$? What could $D^{\frac{1}{2}}$ mean? That question became important to applied mathematicians when such situations began to be encountered in real physical problems.[4]

As this section illustrates, the gamma function lets us answer Euler's interpolation question about the factorial function. What continues to make this interesting, today, is that it is known that there are many ways to define functions that give the correct values of $x!$ for x a positive integer, but for non-integer values of x, these different ways give quite different values for $x!$, none of which agree with the values produced by the gamma function.[5] So, what makes Euler's integral formulation "the right way"? Astonishingly, this question wasn't answered until as recently as 1922.

As the 19th century came to its end, the modern approach to function theory had already moved away from thinking of the equations satisfied by a function as the fundamental concept, and instead believed the geometrical structure of a function was its fundamental property. As Davis observes:

The desired condition was found in notions of convexity. A curve is convex if the following is true of it: take any two points on the curve and join them by a straight line; then the portion of the curve between the points lies

4. For more on this, see my book, *Oliver Heaviside: The Life, Work, and Times of an Electrical Genius of the Victorian Age* (The Johns Hopkins University Press, 2002), in particular, Chapter 10, "Strange Mathematics," pp. 217–240.

5. For more on such alternative functions, see Philip J. Davis, "Leonhard Euler's Integral: A Historical Profile of the Gamma Function," *American Mathematical Monthly*, December 1959, pp. 849–869.

below the line [$f(x) = x^2$, for example, defines a convex curve].[6] A convex curve does not wiggle; it cannot look like a camel's back. At the turn of the century, convexity was in the mathematical air. It was found to be intrinsic to many diverse phenomena. Over the period of a generation, it was sought out, it was generalized, it was abstracted, it was investigated for its own sake, it was applied. Called to attention by the work of H. Brunn in 1887 and of H. Minkowski in 1903 on convex bodies and given an independent interest in 1906 by the work of J.L.W.V. Jensen, the idea of convexity spread and established itself in mean value theory, in potential theory, in topology, and most recently in game theory and linear programming.[7] At the turn of the [20th] century then, an application of convexity to the gamma function would have been natural and in order.[8]

And that's what happened in 1922.

That year it was shown that the gamma function is the only function that has all the properties of being positive for $x > 0$, being equal to 1 at $x = 1$, satisfying the functional equation $\Gamma(x + 1) = x\Gamma(x)$, and being *logarithmically convex*. That last condition means that not only is $\Gamma(x)$ convex but, in addition, $\ln\{\Gamma(x)\}$ is also convex.[9] (You'll recall, from the previous chapter, how $\ln\{\Gamma(x)\}$ arose, in a natural way, in the development of the digamma function.) Somehow, in the mid-1700s, Euler's genius led him to the unique way to develop the logarithmically convex gamma function—150 years before mathematicians began to think of such functions!

6. In most calculus textbooks, such a curve would be called *concave upward* (or *convex downward*). By contrast, $f(x) = \sqrt{x}$ does not define a convex curve.

7. The individuals Davis mentions are the German mathematicians Karl Hermann Brunn (1862–1939) and Einstein's college math professor Hermann Minkowski (1864–1909), and the Dutch mathematician Johan Ludwig William Valdemar Jensen (1859–1925).

8. Davis, "Leonhard Euler's Integral."

9. An example of a logarithmically convex function is $f(x) = e^{x^2}$. An example of a function that is convex but not logarithmically convex is $f(x) = x^2$, as $\ln(x^2) = 2\ln(x)$: joining any two points on the curve of $2\ln(x)$ results in a line that is *below* the function curve. (Sketch it and see!)

Okay, now back to the beta function in (2.2.1). Changing the variable in (1.4.1) to $x = y^2$ (and so $dx = 2y\,dy$), we have

$$\Gamma(n) = \int_0^\infty e^{-y^2} y^{2n-2} 2y\,dy = 2\int_0^\infty e^{-y^2} y^{2n-1}\,dy.$$

We get another true equation if we replace n with m, and the dummy integration variable y with the dummy integration variable x, and so

(2.2.2) $$\Gamma(m) = 2\int_0^\infty e^{-x^2} x^{2m-1}\,dx.$$

Thus,

$$\Gamma(m)\Gamma(n) = 4\int_0^\infty e^{-x^2} x^{2m-1}\,dx \int_0^\infty e^{-y^2} y^{2n-1}\,dy$$

$$= 4\int_0^\infty \int_0^\infty e^{-(x^2+y^2)} x^{2m-1} y^{2n-1}\,dx\,dy.$$

This double integral looks pretty awful, but the trick that brings it to its knees is to switch from Cartesian coordinates to polar coordinates (for a famous application of this trick, see Appendix 2). That is, write $r^2 = x^2 + y^2$, where $x = r\cos(\theta)$ and $y = r\sin(\theta)$, and so the differential area patch $dxdy$ transforms[10] to $rdrd\theta$. When we integrate the double integral over the region $0 \le x, y < \infty$, we are integrating over the entire first quadrant of the plane, which is equivalent to integrating over the region $0 \le r < \infty$, $0 \le \theta \le \frac{\pi}{2}$. So,

$$\Gamma(m)\Gamma(n) = 4\int_0^{\frac{\pi}{2}} \int_0^\infty e^{-r^2} \{r\cos(\theta)\}^{2m-1} \{r\sin(\theta)\}^{2n-1}\ rdrd\theta$$

---

10. The general method for calculating how the differential area patch in a double integral transforms under a change of variables requires the calculation of a determinant that mathematicians call the *Jacobian* (after the Prussian mathematician Carl Jacobi (1804–1851)). This is typically discussed in an advanced calculus course (certainly it is beyond AP-calculus), and so here I am depending on your knowledge from high school trigonometry and geometry about how Cartesian and polar coordinates, in particular, are related.

or

$$(2.2.3) \quad \Gamma(m)\Gamma(n) = \left[2\int_0^\infty e^{-r^2} r^{2(m+n)-1} dr \right] \left[2\int_0^{\frac{\pi}{2}} \cos^{2m-1}(\theta)\sin^{2n-1}(\theta)d\theta \right].$$

Let's now examine, in turn, each of the integrals in square brackets on the right in (2.2.3).

First, if you compare

$$2\int_0^\infty e^{-r^2} r^{2(m+n)-1} dr$$

to (2.2.2), you see that they are the same if we associate $x \leftrightarrow r$ and $m \leftrightarrow (m + n)$. Making those replacements, the first square-bracket term in (2.2.3) becomes

$$2\int_0^\infty e^{-r^2} r^{2(m+n)-1} dr = \Gamma(m+n).$$

Thus,

$$(2.2.4) \quad \Gamma(m)\Gamma(n) = \Gamma(m+n)\left[2\int_0^{\frac{\pi}{2}} \cos^{2m-1}(\theta)\sin^{2n-1}(\theta)d\theta \right].$$

Next, returning to (2.2.1), the definition of the beta function, make the change of variable $x = \cos^2(\theta)$ (and so $dx = -2\sin(\theta)\cos(\theta)d\theta$), which says that $1 - x = \sin^2(\theta)$. So,

$$B(m,n) = \int_0^1 x^{m-1}(1-x)^{n-1}\, dx$$

$$= -2\int_{\frac{\pi}{2}}^0 \cos^{2m-2}(\theta)\sin^{2n-2}(\theta)\sin(\theta)\cos(\theta)d\theta$$

or

$$B(m,n) = 2\int_0^{\frac{\pi}{2}} \cos^{2m-1}(\theta)\sin^{2n-1}(\theta)d\theta,$$

which is the integral in the square brackets of (2.2.4). Therefore,

$$\Gamma(m)\,\Gamma(n) = \Gamma(m+n)\,B(m,n)$$

or, rewriting, we have a very important result, one that ties the gamma and beta functions together:

(2.2.5)
$$B(m,n) = \frac{\Gamma(m)\Gamma(n)}{\Gamma(m+n)}.$$

There is a famous result that we won't need until later in the book, but we can derive it right now with the aid of the beta function. We start with (2.2.5),

$$B(m,n) = \frac{\Gamma(m)\Gamma(n)}{\Gamma(m+n)} = \frac{(m-1)!(n-1)!}{(m+n-1)!},$$

from which it follows that

$$B(m+1,n+1) = \frac{m!n!}{(m+n+1)!}.$$

So, writing $m = n = z$, we have

$$B(z+1,z+1) = \frac{z!z!}{(2z+1)!}.$$

From the definition of the beta function in (2.2.1), we have

$$B(z+1,z+1) = \int_0^1 x^z (1-x)^z \, dx$$

and so

$$\frac{z!z!}{(2z+1)!} = \int_0^1 x^z (1-x)^z \, dx.$$

Next, make the change of variable $x = \frac{1+s}{2}$ (and so $1-x = \frac{1-s}{2}$) to get

$$\frac{z!z!}{(2z+1)!} = \int_{-1}^1 \left(\frac{1+s}{2}\right)^z \left(\frac{1-s}{2}\right)^z \frac{1}{2} \, ds$$

$$= 2^{-2z-1} \int_{-1}^1 (1-s^2)^z \, ds$$

or, since the integrand is even,

$$\frac{z!z!}{(2z+1)!} = 2^{-2z} \int_0^1 (1-s^2)^z \, ds.$$

Make a second change of variable now, to $u = s^2$ (and so $ds = \frac{du}{2\sqrt{u}}$), to arrive at

$$\frac{z!z!}{(2z+1)!} = 2^{-2z} \int_0^1 (1-u)^z \frac{du}{2\sqrt{u}} = 2^{-2z-1} \int_0^1 (1-u)^z u^{-\frac{1}{2}} \, du.$$

The last integral is, from (2.2.1),

$$B\left(\frac{1}{2}, z+1\right) = \frac{\Gamma\left(\frac{1}{2}\right)\Gamma(z+1)}{\Gamma\left(z+\frac{3}{2}\right)} = \frac{\left(-\frac{1}{2}\right)! z!}{\left(z+\frac{1}{2}\right)!}$$

and so, recalling from (1.4.6) that $(-\frac{1}{2})! = \sqrt{\pi}$, we have

$$\frac{z!z!}{(2z+1)!} = 2^{-2z-1} \frac{z!\sqrt{\pi}}{\left(z+\dfrac{1}{2}\right)!}.$$

Canceling a $z!$ on each side, and then cross-multiplying, gives us

(2.2.6) $$z!\left(z+\frac{1}{2}\right)! = 2^{-2z-1}\sqrt{\pi}\,(2z+1)!$$

and since

$$\left(z+\frac{1}{2}\right)! = \left(z+\frac{1}{2}\right)\left(z-\frac{1}{2}\right)! = \left(\frac{2z+1}{2}\right)\left(z-\frac{1}{2}\right)!$$

and

$$(2z+1)! = (2z+1)(2z)!,$$

we can then alternatively write

(2.2.7) $$z!\left(z-\frac{1}{2}\right)! = 2^{-2z}\sqrt{\pi}\,(2z)!$$

(2.2.6) and (2.2.7) are variations on what mathematicians commonly call the *Legendre duplication formula.*[11] We'll use (2.2.7), in particular, when we derive what is called the *functional equation* of the zeta function.

The duplication formula is very useful for (among other things) calculating the factorial function of half-integer values. For example, to find $\left(5\frac{1}{2}\right)!$, set $z = 6$ in (2.2.7) to write

$$6!\left(5\frac{1}{2}\right)! = 2^{-12}\sqrt{\pi}(12!)$$

11. Named after, as you no doubt suspect, Adrien-Marie Legendre (see note 4 in Chapter 1), who discovered (2.2.6)/(2.2.7) in 1809.

and so

$$\left(5\frac{1}{2}\right)! = \sqrt{\pi}\,\frac{12!}{2^{12}6!} = \sqrt{\pi}\,\frac{(12)(11)(10)(9)(8)(7)}{4,096} = 287.885277\ldots.$$

While handy for carrying out this task, it's worth noting that we could still do the calculation even if we didn't have the duplication formula. That's because we know $(n+1)! = (n+1)n!$. So, as an alternative way to calculate $(5\frac{1}{2})!$, we set $n = 4\frac{1}{2}$ and write

$$\left(5\frac{1}{2}\right)! = \left(4\frac{1}{2}+1\right)\left(4\frac{1}{2}\right)! = \frac{11}{2}\left(4\frac{1}{2}\right)!$$

and then, in the same way, continue by writing

$$\left(4\frac{1}{2}\right)! = \left(3\frac{1}{2}+1\right)\left(3\frac{1}{2}\right)! = \frac{9}{2}\left(3\frac{1}{2}\right)!,$$

$$\left(3\frac{1}{2}\right)! = \left(2\frac{1}{2}+1\right)\left(2\frac{1}{2}\right)! = \frac{7}{2}\left(2\frac{1}{2}\right)!,$$

$$\left(2\frac{1}{2}\right)! = \left(1\frac{1}{2}+1\right)\left(1\frac{1}{2}\right)! = \frac{5}{2}\left(1\frac{1}{2}\right)!,$$

$$\left(1\frac{1}{2}\right)! = \left(\frac{1}{2}+1\right)\left(\frac{1}{2}\right)! = \frac{3}{2}\left(\frac{1}{2}\right)!.$$

Since we know from the last chapter that $(\frac{1}{2})! = \frac{1}{2}\sqrt{\pi}$, then

$$\left(5\frac{1}{2}\right)! = \left(\frac{11}{2}\right)\left(\frac{9}{2}\right)\left(\frac{7}{2}\right)\left(\frac{5}{2}\right)\left(\frac{3}{2}\right)\left(\frac{1}{2}\right)!$$

$$= \frac{10,395}{32}\left(\frac{1}{2}\sqrt{\pi}\right) = 287.885277\ldots,$$

just as we got from the duplication formula.

Challenge Problem 2.2.1: Using the same approach just described, calculate the value of $(-5\tfrac{1}{2})$! Hint: It's not very large.

2.3 Euler Almost Computes ζ(3)

Just after the third box in Section 1.2, I promised you a derivation of yet another result originally due to Euler, a result that bears a superficial resemblance to $\zeta(3)$:

$$1-\frac{1}{3^3}+\frac{1}{5^3}-\frac{1}{7^3}+\cdots=\frac{\pi^3}{32}.$$

Where does this come from? It all started with a 1744 letter to a friend. In that letter, Euler claimed that

$$(2.3.1) \qquad \frac{\pi-t}{2}=\sin(t)+\frac{\sin(2t)}{2}+\frac{\sin(3t)}{3}+\frac{\sin(4t)}{4}+\frac{\sin(5t)}{5}+\cdots$$

$$=\sum_{n=1}^{\infty}\frac{\sin(nt)}{n},$$

a claim we'll establish in the next chapter through the use of Fourier series, a mathematical theory named after the French mathematical physicist Joseph Fourier (1768–1830). Of course, (2.3.1) can't be true for arbitrary values of t (for $t = 0$, for example, the claim would be $\frac{\pi}{2} = 0$, which seems unlikely). It *is* true, however, for all t in the open interval $0 < t < 2\pi$ (*open* means the endpoint values of 0 and 2π are excluded). For now, we'll just assume Euler was correct and that (2.3.1) is indeed true.

One point that may be puzzling you, however, is how did *Euler* come up with (2.3.1), decades before Fourier was born, and even more decades before Fourier's masterpiece on Fourier series (*The Analytical Theory of Heat*[12]) was published in 1822? Euler didn't have

---

12. Fourier's book is all about how to solve the second-order partial differential equation that mathematical physicists call the *heat equation*, about which you can read much more in my book, *Hot Molecules, Cold Electrons* (Princeton University Press, 2020).

the theory of Fourier series to work with—so just what *did* he do to get (2.3.1)? The answer is that he created a typical Eulerian conjuring derivation seemingly out of thin air, in an analysis that is borderline crazy—but it works! I won't discuss it here, because the Fourier series derivation in the next chapter is the proper way to get (2.3.1), in a routine way.[13]

To get Euler's "near-miss" of $\zeta(3)$, we start by integrating (2.3.1) over the interval $0 < t < x$:

$$\int_0^x \left\{ \frac{\pi - t}{2} \right\} dt = \int_0^x \left\{ \sum_{n=1}^{\infty} \frac{\sin(nt)}{n} \right\} dt.$$

That is,

$$\frac{\pi}{2} x - \frac{x^2}{4} = \sum_{n=1}^{\infty} \frac{1}{n} \int_0^x \sin(nt) dt = \sum_{n=1}^{\infty} \frac{1}{n} \left\{ -\frac{\cos(nt)}{n} \right\} \Big|_0^x$$

$$= \sum_{n=1}^{\infty} \frac{1 - \cos(nx)}{n^2}$$

or

$$\frac{\pi}{2} x - \frac{x^2}{4} = \sum_{n=1}^{\infty} \frac{1}{n^2} - \sum_{n=1}^{\infty} \frac{\cos(nx)}{n^2}.$$

The first sum on the right is, of course, $\zeta(2)$, and so we have

(2.3.2) $$\sum_{n=1}^{\infty} \frac{\cos(nx)}{n^2} = \frac{\pi^2}{6} - \frac{\pi}{2} x + \frac{x^2}{4} = \frac{3x^2 - 6\pi x + 2\pi^2}{12}.$$

Next, integrate (2.3.2) over the interval $0 < x < u$:

$$\int_0^u \sum_{n=1}^{\infty} \frac{\cos(nx)}{n^2} dx = \int_0^u \frac{3x^2 - 6\pi x + 2\pi^2}{12} dx$$

---

13. But if you're curious, you can find Euler's wild derivation in my book, *Dr. Euler's Fabulous Formula* (Princeton University Press, 2006), pp. 134–136.

96

Chapter 2

or

$$\sum_{n=1}^{\infty}\frac{1}{n^2}\int_0^u \cos(nx)dx = \left(\frac{x^3}{12}-\frac{\pi x^2}{4}+\frac{\pi^2 x}{6}\right)\Big|_0^u$$

and so

(2.3.3) $$\sum_{n=1}^{\infty}\frac{1}{n^2}\left\{\frac{\sin(nx)}{n}\right\}\Big|_0^u=\sum_{n=1}^{\infty}\frac{\sin(nu)}{n^3}=\frac{u^3}{12}-\frac{\pi u^2}{4}+\frac{\pi^2 u}{6}.$$

If we set $u=\frac{\pi}{2}$, we have

$$\sum_{n=1}^{\infty}\frac{\sin\left(\frac{n\pi}{2}\right)}{n^3}=1-\frac{1}{3^3}+\frac{1}{5^3}-\frac{1}{7^3}+\cdots=\frac{\pi^3}{96}-\frac{\pi^3}{16}+\frac{\pi^3}{12}=\frac{3\pi^3}{96}$$

or

(2.3.4) $$1-\frac{1}{3^3}+\frac{1}{5^3}-\frac{1}{7^3}+\cdots=\frac{\pi^3}{32}$$

which is the formula that opens this section. It's unquestionably a beautiful formula, yes, but it isn't $\zeta(3)$. At best, it's $\zeta(3)$'s third cousin. Your disappointment (and mine) in that is almost surely just a small fraction of what must have been Euler's.

We can't leave (2.3.4) without admitting that it is simply impossible to resist the urge to calculate both sides of Euler's "near-miss" formula. The right-hand side is

$$\frac{\pi^3}{32}=0.9689461462593\ldots,$$

while summing the first 10,000 terms of the alternating series on the left gives 0.9689461462593 . . . , and we see agreement out to an impressive 13 decimal places. This is, of course, not a mathematical proof by any means, but if the formula is wrong, then we've just witnessed an incredibly amazing coincidence of astonishing improbability.

Challenge Problem 2.3.1: Derive these two formulas:

$$\frac{1}{2^2} + \frac{1}{4^2} + \frac{1}{6^2} + \frac{1}{8^2} + \cdots = \frac{1}{4}\zeta(2)$$

and

$$1 - \frac{1}{2^2} + \frac{1}{3^2} - \frac{1}{4^2} + \frac{1}{5^2} - \cdots = \frac{1}{2}\zeta(2).$$

2.4 Integral Forms of $\zeta(2)$ and $\zeta(3)$

In this section, we'll come at the zeta function in a different way, using just integrals (no sums). In fact, I'll show you two quite different such analyses, each relatively modern (from opposite ends of the 20th century). My first example begins by reminding you of a remark I made in Chapter 1 (Section 1.2), concerning the irrationality of $\zeta(3)$. As I stated there, in 1979, the French mathematician Roger Apéry showed that $\zeta(3)$ is irrational, a result rightfully considered to be a tremendous achievement.[14] That's because, before Apéry, every mathematician on the planet would probably have bet a week's salary (or $100, whichever was smaller) that $\zeta(3)$ is indeed irrational—but nobody could prove it. Then, suddenly, Apéry did it, and justifiably became famous (in the world of mathematics, anyway). Then, that same year, the Dutch mathematician Frits Beukers (born 1953) published a much simpler proof, thus illustrating, like the four-minute mile, that once somebody does what up to then had been thought to be extraordinarily difficult (if not simply impossible), then it's more than likely that everybody will start doing it![15]

14. R. Apéry, "Irrationalité de $\zeta(2)$ and $\zeta(3)$," *Astérisque*, vol. 61, 1979, pp. 11–13.

15. F. Beukers, "A Note on the Irrationality of $\zeta(2)$ and $\zeta(3)$," *Bulletin of the London Mathematical Society*, vol. 11, 1979, pp. 268–272.

Beukers' work was based on the perhaps benign-looking double integral

$$\int_0^1 \int_0^1 \frac{dxdy}{1-xy},$$

which appears without explanation at the beginning of his paper. It is now, in fact, often called *Beukers' integral*. Despite that, this double integral has a history that reaches back more than 70 years before Beukers, to a high school math teacher. The story begins in 1908 when Paul Stäckel (1862–1919), a German university math professor, published a brief note, in which he observed that the above double integral over the unit square is equal to $\sum_{n=1}^{\infty} \frac{1}{n^2}$, a sum known since 1734 when Euler showed it to have the value $\frac{\pi^2}{6}$. It must therefore be true that

$$\int_0^1 \int_0^1 \frac{dxdy}{1-xy} = \frac{\pi^2}{6}.$$

Stäckel then asked for a direct evaluation of the double integral, that is, for an evaluation that avoids any reference to, and is independent of, Euler's classic result. Stäckel died young, of a brain tumor, but he lived long enough to see his challenge answered.

That success came in 1913, when a German high school math teacher in Berlin, Franz Goldscheider (1852–1926), published an almost equally brief note in which he evaluated Stäckel's double integral through the use of an enormously clever sequence of changes of variables. It was a tour de force derivation that Euler, a master himself of devilishly ingenious symbolic manipulations that appear to come from seemingly out of nowhere, would have loved.[16]

First, however, let's quickly establish the equality of the double integral with the original infinite series for $\zeta(2)$. As Stäckel wrote at the beginning of his note,

16. For those readers who want to check the accuracy of my reading of German, the notes of Stäckel and Goldscheider both appeared in *Archiv der Mathematik und Physik*, vol. 13, 1908, p. 362, and vol. 20, 1913, pp. 323–324, respectively.

$$\frac{1}{1-xy} = 1 + xy + (xy)^2 + (xy)^3 + (xy)^4 + \cdots, -1 < xy < 1.$$

Thus,

$$\int_0^1 \int_0^1 \frac{dxdy}{1-xy} = \int_0^1 \int_0^1 \{1 + xy + x^2 y^2 + x^3 y^3 + x^4 y^4 + \cdots\} dxdy$$

$$= \int_0^1 \left\{ x + \frac{1}{2} x^2 y + \frac{1}{3} x^3 y^2 + \frac{1}{4} x^4 y^3 + \frac{1}{5} x^5 y^4 + \cdots \right\} |_0^1 \, dy$$

$$= \int_0^1 \left\{ 1 + \frac{1}{2} y + \frac{1}{3} y^2 + \frac{1}{4} y^3 + \frac{1}{5} y^4 + \cdots \right\} dy$$

$$= \left\{ y + \frac{1}{2}\left(\frac{1}{2} y^2\right) + \frac{1}{3}\left(\frac{1}{3} y^3\right) + \frac{1}{4}\left(\frac{1}{4} y^4\right) + \frac{1}{5}\left(\frac{1}{5} y^5\right) + \cdots \right\} |_0^1$$

$$= 1 + \frac{1}{2^2} + \frac{1}{3^2} + \frac{1}{4^2} + \frac{1}{5^2} + \cdots = \sum_{n=1}^{\infty} \frac{1}{n^2} = \zeta(2).$$

Now, starting with Stäckel's integral (called P), Goldscheider introduced a second, similar double integral that he called Q:

(2.4.1)
$$P = \int_0^1 \int_0^1 \frac{dxdy}{1-xy}$$

and

(2.4.2)
$$Q = \int_0^1 \int_0^1 \frac{dxdy}{1+xy}.$$

He then formed $P - Q$ to get

(2.4.3) $P - Q = \int_0^1 \int_0^1 \left\{ \dfrac{1}{1-xy} - \dfrac{1}{1+xy} \right\} dx\,dy = \int_0^1 \int_0^1 \dfrac{2xy}{1-x^2 y^2}\,dx\,dy.$

Changing variables to $x^2 = u$ and $y^2 = v$ (and so $dx = \frac{du}{2x}$ and $dy = \frac{dv}{2y}$), he arrived at

$$dx\,dy = \frac{du\,dv}{4xy}$$

and so, since u and v also each vary from 0 to 1, (2.4.3) becomes

$$P - Q = \frac{1}{2} \int_0^1 \int_0^1 \frac{du\,dv}{1-uv} = \frac{1}{2} P,$$

from which it quickly follows that

(2.4.4) $P = 2Q.$

Next, Goldscheider changed variable in P to $u = -y$ (and so $dy = -du$, with u varying from 0 to -1 as y varies from 0 to 1) to get

$$P = \int_0^1 \int_0^{-1} \frac{dx(-du)}{1+xu} = \int_0^1 \int_{-1}^0 \frac{dx\,du}{1+xu} = \int_0^1 \int_{-1}^0 \frac{dx\,dy}{1+xy}.$$

Thus, forming $P + Q$, he had

$$P + Q = \int_0^1 \int_{-1}^0 \frac{dx\,dy}{1+xy} + \int_0^1 \int_0^1 \frac{dx\,dy}{1+xy} = \int_0^1 \int_{-1}^1 \frac{dx\,dy}{1+xy}$$

or

(2.4.5) $P + Q = \int_{-1}^1 dy \left\{ \int_0^1 \dfrac{dx}{1+xy} \right\}.$

He then made yet another change of variables, to $u = y + \frac{1}{2} x(y^2 - 1)$, and so as y varies from -1 to 1 so does u. Since

$$\frac{du}{dy} = 1 + xy \left(\text{and so } dy = \frac{du}{1+xy} \right),$$

he therefore saw that (2.4.5) becomes

$$(2.4.6) \quad P + Q = \int_{-1}^{1} du \left\{ \int_{0}^{1} \frac{dx}{(1+xy)^2} \right\} = \int_{-1}^{1} du \left\{ \int_{0}^{1} \frac{dx}{1 + 2xy + x^2 y^2} \right\}.$$

As unpromising as (2.4.6) might appear to be at first glance (those y's in the denominator of the integrand look troublesome), it actually isn't a disaster. That's because

$$1 + 2ux + x^2 = 1 + 2 \left[y + \frac{1}{2} x(y^2 - 1) \right] x + x^2 = 1 + 2xy + x^2 y^2$$

and so (2.4.6) becomes

$$(2.4.7) \qquad P + Q = \int_{-1}^{1} du \left\{ \int_{0}^{1} \frac{dx}{1 + 2ux + x^2} \right\}.$$

Well, you might say in response, (2.4.7) still doesn't really look all that terrific, either.

Have faith! Goldscheider saved the day by pulling a final change of variable out of his hat, with $u = \cos(\varphi)$ (and so $du = -\sin(\varphi)d\varphi$). As u varies from -1 to 1, we see φ varying from π to 0, and so (2.4.7) becomes

$$(2.4.8) \qquad P + Q = \int_{\pi}^{0} d\varphi \left\{ \int_{0}^{1} \frac{-\sin(\varphi)dx}{1 + 2x\cos(\varphi) + x^2} \right\}$$

$$= \int_{0}^{\pi} d\varphi \left\{ \int_{0}^{1} \frac{\sin(\varphi)dx}{1 + 2x\cos(\varphi) + x^2} \right\}.$$

We can avoid having (2.4.8) make us think we've driven over the edge of a cliff by recalling the classic differentiation formula

$$\frac{d}{d\theta}\tan^{-1}(s)=\left(\frac{1}{1+s^2}\right)\frac{ds}{d\theta}.$$

If we apply this formula to the case of

$$s=\frac{x+\cos(\varphi)}{\sin(\varphi)}, \theta=x,$$

we have

$$\frac{d}{dx}\left[\tan^{-1}\left\{\frac{x+\cos(\varphi)}{\sin(\varphi)}\right\}\right]=\left(\frac{1}{1+\left\{\dfrac{x+\cos(\varphi)}{\sin(\varphi)}\right\}^2}\right)\frac{1}{\sin(\varphi)}$$

$$=\frac{\sin(\varphi)}{1+2x\cos(\varphi)+x^2}.$$

Putting this into (2.4.8), we have

$$P+Q=\int_0^\pi d\varphi\left(\int_0^1 \frac{d}{dx}\left[\tan^{-1}\left\{\frac{x+\cos(\varphi)}{\sin(\varphi)}\right\}\right]dx\right)$$

$$=\int_0^\pi d\varphi\left(\int_0^1 d\left[\tan^{-1}\left\{\frac{x+\cos(\varphi)}{\sin(\varphi)}\right\}\right]\right)$$

$$=\int_0^\pi d\varphi\left(\tan^{-1}\left\{\frac{x+\cos(\varphi)}{\sin(\varphi)}\right\}\Big|_{x=0}^{x=1}\right)$$

or

(2.4.9) $$P+Q=\int_0^\pi\left(\tan^{-1}\left\{\frac{1+\cos(\varphi)}{\sin(\varphi)}\right\}-\tan^{-1}\left\{\frac{\cos(\varphi)}{\sin(\varphi)}\right\}\right)d\varphi.$$

The integral in (2.4.9) may look like a tiger, but it's actually a pussycat. That's because, first of all,

$$\tan^{-1}\left\{\frac{\cos(\varphi)}{\sin(\varphi)}\right\} = \tan^{-1}\left\{\frac{1}{\tan(\varphi)}\right\} = \frac{\pi}{2} - \varphi,$$

a result that follows by simply drawing an arbitrary right triangle and observing that the tangents of the triangle's two acute angles are reciprocals of each other. If φ is one of those angles, then $\frac{\pi}{2} - \varphi$ is the other one. And second, after recalling a couple of identities from trigonometry, we have from a similar line of reasoning that

$$\tan^{-1}\left\{\frac{1+\cos(\varphi)}{\sin(\varphi)}\right\} = \tan^{-1}\left\{\frac{2\cos^2\left(\frac{\varphi}{2}\right)}{2\sin\left(\frac{\varphi}{2}\right)\cos\left(\frac{\varphi}{2}\right)}\right\}$$

$$= \tan^{-1}\left\{\frac{1}{\tan\left(\frac{\varphi}{2}\right)}\right\} = \frac{\pi}{2} - \frac{\varphi}{2}.$$

Thus, (2.4.9) becomes

$$P + Q = \int_0^\pi \left[\left(\frac{\pi}{2} - \frac{\varphi}{2}\right) - \left(\frac{\pi}{2} - \varphi\right)\right]d\varphi = \frac{1}{2}\int_0^\pi \varphi \, d\varphi = \frac{\pi^2}{4}.$$

Since $P = 2Q$ from (2.4.4), it then immediately follows that $Q = \frac{\pi^2}{12}$ and so, just like that,

$$P = \int_0^1\int_0^1 \frac{dxdy}{1-xy} = \frac{\pi^2}{6} = \zeta(2),$$

and Professor Stäckel had the derivation he had requested five years earlier.[17]

Goldscheider's derivation evaluates P in a somewhat roundabout manner, in conjunction with a second double integral (Q). Decades later, the Greek-born American mathematician Tom Apostol (1923–2016) discussed a markedly different direct evaluation of Stäckel's double integral P.[18] Of this new approach, he wrote "This evaluation has been presented [by Apostol] for a number of years in elementary calculus courses [at Caltech] but does not seem to be recorded in the literature." Apostol gives no details in his paper on how *he* came to know the analysis he describes, but the strong implication is that it was not original with him (and, so, we have yet another historical puzzle for math aficionados to pursue).

To start Apostol's analysis, we first rotate the x, y coordinate axes in such a way as to transform the unit square (the region of integration for P) into a region (one that is still square) with more symmetry about the axes. Just to remind you of the classic axes-rotation equations (which you can find derived in any good high school book on analytic geometry), if a point has coordinates x, y in the original system, then a counterclockwise rotation of the axes through the angle a results in the following relationships between the coordinates x', y' of the point in the new system and x, y:

$$x = x' \cos(a) - y' \sin(a), y = x' \sin(a) + y' \cos(a).$$

Using $a = 45°$ (and writing u and v for x' and y', respectively, to use Apostol's notation), we arrive at

(2.4.10) $$x = u\frac{1}{\sqrt{2}} - v\frac{1}{\sqrt{2}} = \frac{u-v}{\sqrt{2}}$$

17. It is easy to show that $\zeta(3) = \sum_{n=1}^{\infty} \frac{1}{n^3} = \int_0^1 \int_0^1 \int_0^1 \frac{dxdydz}{1-xyz}$, but a direct evaluation of this triple integral over the unit cube continues to elude all who have tried.

18. T. M. Apostol, "A Proof That Euler Missed: Evaluating $\zeta(2)$ the Easy Way," *Mathematical Intelligencer* 1983 (no. 3), pp. 59–60.

and

(2.4.11)
$$y = u\frac{1}{\sqrt{2}} + v\frac{1}{\sqrt{2}} = \frac{u+v}{\sqrt{2}}.$$

The particular rotation angle of 45° puts the new horizontal axis (u) along the diagonal of the square region of integration, as shown in Figure 2.4.1. (Remember, we rotated the axes counterclockwise, and so the square appears rotated clockwise.)

Next, we write

(2.4.12)
$$\int_0^1\int_0^1 \frac{dx\,dy}{1-xy} = \int_0^1\int_0^1 \frac{dA_{x,y}}{1-xy},$$

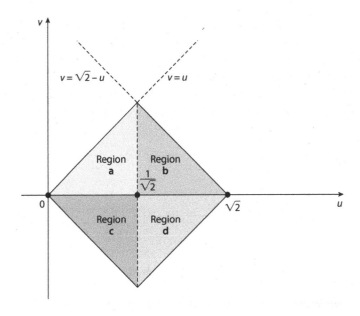

FIGURE 2.4.1.

The unit square after a 45° counterclockwise rotation of the axes.

where the differential area patch in the x, y system is $dA_{x,y} = dxdy$. The differential area patch in the u, v system is, by analogy, $dA_{u,v} = dudv$. Since the value of the integral is independent of the coordinate system, and since (2.4.10) and (2.4.11) tell us that

$$1 - xy = 1 - \frac{u^2 - v^2}{2} = \frac{2 - u^2 + v^2}{2},$$

then

(2.4.13)
$$\int_0^1 \int_0^1 \frac{dxdy}{1 - xy} = \iint_{\text{rotated square}} \frac{dA_{u,v}}{\dfrac{2 - u^2 + v^2}{2}}$$

$$= 2\iint_{\text{rotated square}} \frac{dudv}{2 - u^2 + v^2}.$$

This formulation of the problem is useful because of the symmetry of the rotated square about the u-axis, as well as the fact that v appears as squared in the integrand. This means that the contribution to the integral from region **a** above the horizontal axis is equal to the contribution from region **c** below the horizontal axis, and similarly for the contributions of regions **b** and **d**. Thus,

$$\int_0^1 \int_0^1 \frac{dxdy}{1 - xy} = 2\left[2\iint_{\text{region } a} \frac{1}{2 - u^2 + v^2} dudv + 2\iint_{\text{region } b} \frac{1}{2 - u^2 + v^2} dudv \right]$$

$$= 4\int_0^{1/\sqrt{2}} \left\{ \int_0^u \frac{dv}{2 - u^2 + v^2} \right\} du + 4\int_{1/\sqrt{2}}^{\sqrt{2}} \left\{ \int_0^{\sqrt{2} - u} \frac{dv}{2 - u^2 + v^2} \right\} du,$$

where we imagine each double integral is the sum of the areas of an infinite number of vertical strips, each of differential width du. Both of the final two inner integrals (with respect to v) are easy to do.

Recalling the fundamental result

$$\int_0^x \frac{dt}{a^2+t^2} = \frac{1}{a}\tan^{-1}\left(\frac{x}{a}\right),$$

we then have

$$\int_0^u \frac{dv}{2-u^2+v^2} = \int_0^u \frac{dv}{(\sqrt{2-u^2})^2+v^2} = \frac{1}{\sqrt{2-u^2}}\tan^{-1}\left(\frac{u}{\sqrt{2-u^2}}\right)$$

and

$$\int_0^{\sqrt{2}-u} \frac{dv}{2-u^2+v^2} = \int_0^{\sqrt{2}-u} \frac{dv}{(\sqrt{2-u^2})^2+v^2} = \frac{1}{\sqrt{2-u^2}}\tan^{-1}\left(\frac{\sqrt{2}-u}{\sqrt{2-u^2}}\right).$$

Thus,

$$\int_0^1\int_0^1 \frac{dxdy}{1-xy} = 4\int_0^{1/\sqrt{2}} \tan^{-1}\left(\frac{u}{\sqrt{2-u^2}}\right)\frac{du}{\sqrt{2-u^2}}$$

$$+ 4\int_{1/\sqrt{2}}^{\sqrt{2}} \tan^{-1}\left(\frac{\sqrt{2}-u}{\sqrt{2-u^2}}\right)\frac{du}{\sqrt{2-u^2}}$$

or

(2.4.14) $$\int_0^1\int_0^1 \frac{dxdy}{1-xy} = I_1 + I_2,$$

where

(2.4.15) $$I_1 = 4\int_0^{1/\sqrt{2}} \tan^{-1}\left(\frac{u}{\sqrt{2-u^2}}\right)\frac{du}{\sqrt{2-u^2}}$$

and

$$(2.4.16) \qquad I_2 = 4\int_{1/\sqrt{2}}^{\sqrt{2}} \tan^{-1}\left(\frac{\sqrt{2}-u}{\sqrt{2-u^2}}\right)\frac{du}{\sqrt{2-u^2}}.$$

It turns out, fortunately, that I_1 and I_2 are also not difficult to do.

For I_1: Let $u = \sqrt{2}\sin(\theta)$ and so

$$du = \sqrt{2}\cos(\theta)d\theta = \sqrt{2}\sqrt{1-\sin^2(\theta)} = \sqrt{2}\sqrt{1-\frac{u^2}{2}}d\theta = \sqrt{2}\frac{\sqrt{2-u^2}}{\sqrt{2}}d\theta$$

or

$$d\theta = \frac{du}{\sqrt{2-u^2}}.$$

Also,

$$\tan^{-1}\left\{\frac{u}{\sqrt{2-u^2}}\right\} = \tan^{-1}\left\{\frac{\sqrt{2}\sin(\theta)}{\sqrt{2-2\sin^2(\theta)}}\right\}$$

$$= \tan^{-1}\left\{\frac{\sqrt{2}\sin(\theta)}{\sqrt{2}\cos(\theta)}\right\} = \tan^{-1}\{\tan(\theta)\} = \theta.$$

So

$$I_1 = 4\int_0^{1/\sqrt{2}} \tan^{-1}\left(\frac{u}{\sqrt{2-u^2}}\right)\frac{du}{\sqrt{2-u^2}}$$

$$= 4\int_0^{\pi/6}\theta d\theta = 4\left(\frac{1}{2}\theta^2\right)\Big|_0^{\pi/6} = 2\left(\frac{\pi}{6}\right)^2.$$

For I_2: Let $u = \sqrt{2}\cos(2\theta)$ and so

$$\frac{du}{d\theta} = -(\sqrt{2})2\sin(2\theta)$$

or

$$du = -2\sqrt{2}\sqrt{1-\cos^2(2\theta)}d\theta = -2\sqrt{2}\sqrt{1-\frac{u^2}{2}}d\theta$$

or

$$2d\theta = -\frac{du}{\sqrt{2-u^2}}.$$

Also,

$$\tan^{-1}\left(\frac{\sqrt{2}-u}{\sqrt{2-u^2}}\right) = \tan^{-1}\left\{\frac{\sqrt{2}-\sqrt{2}\cos(2\theta)}{\sqrt{2-2\cos^2(2\theta)}}\right\}$$

$$= \tan^{-1}\left\{\frac{\sqrt{2}(1-\cos(2\theta))}{\sqrt{2}\sqrt{1-\cos^2(2\theta)}}\right\}$$

$$= \tan^{-1}\left\{\frac{\sqrt{1-\cos(2\theta)}}{\sqrt{1+\cos(2\theta)}}\right\} = \tan^{-1}\left\{\sqrt{\frac{2\sin^2(\theta)}{2\cos^2(\theta)}}\right\}$$

$$= \tan^{-1}\{\tan(\theta)\} = \theta.$$

So,

$$I_2 = 4\int_{1/\sqrt{2}}^{\sqrt{2}} \tan^{-1}\left(\frac{\sqrt{2}-u}{\sqrt{2-u^2}}\right)\frac{du}{\sqrt{2-u^2}}$$

$$= -8\int_{\pi/6}^{0}\theta d\theta = 8\left(\frac{1}{2}\theta^2\right)\Big|_0^{\pi/6} = 4\left(\frac{\pi}{6}\right)^2.$$

Thus, and finally,

$$\int_0^1\int_0^1\frac{dxdy}{1-xy} = 2\left(\frac{\pi}{6}\right)^2 + 4\left(\frac{\pi}{6}\right)^2 = 6\left(\frac{\pi}{6}\right)^2 = \frac{\pi^2}{6} = \zeta(2).$$

Following Beukers' success with his proof of the irrationality of $\zeta(3)$ using the P integral of Stäckel and Goldscheider, a cottage industry in evaluating similar-looking integrals has developed. Here is an example: For n any positive integer, what is the value of

$$\int_0^1 \int_0^1 \frac{x^n y^n \ln(xy)}{1-xy} \, dx \, dy \; ?$$

This probably looks pretty scary, but it has the following beautiful solution.

We start by defining, for some parameter σ,

$$(2.4.17) \quad I(\sigma) = \int_0^1 \int_0^1 \frac{(xy)^{n+\sigma}}{1-xy} \, dx \, dy = \int_0^1 \int_0^1 \frac{(xy)^n (xy)^\sigma}{1-xy} \, dx \, dy$$

$$= \int_0^1 \int_0^1 \frac{(xy)^n \, e^{\ln(xy)^\sigma}}{1-xy} \, dx \, dy = \int_0^1 \int_0^1 \frac{(xy)^n \, e^{\sigma \ln(xy)}}{1-xy} \, dx \, dy$$

from which it immediately follows that

$$(2.4.18) \qquad \frac{dI}{d\sigma} = \int_0^1 \int_0^1 \frac{(xy)^n \, \ln(xy) e^{\sigma \ln(xy)}}{1-xy} \, dx \, dy,$$

assuming that the derivative of the integral is the integral of the derivative (of the integrand).[19] If we set $\sigma = 0$, then (2.4.18) says

$$\frac{dI}{d\sigma}\Big|_{\sigma=0} = \int_0^1 \int_0^1 \frac{(xy)^n \, \ln(xy)}{1-xy} \, dx \, dy,$$

19. The operations of integration and differentiation are both defined as *limit* operations, and so what we are assuming is that we can reverse the order of two limit operations. This is often okay, but it's not always okay. A more careful analysis than I'm doing here would spend some time justifying doing the reversal.

and so our plan of attack is now clear. We'll first calculate $I(\sigma)$ as defined in (2.4.17), then we'll differentiate the result with respect to σ, and finally we'll set $\sigma = 0$.

Expanding the integrand of (2.4.17), we have

$$\frac{(xy)^{n+\sigma}}{1-xy} = (xy)^{n+\sigma}[1 + xy + x^2y^2 + x^3y^3 + \cdots]$$

$$= x^{n+\sigma}y^{n+\sigma} + x^{n+1+\sigma}y^{n+1+\sigma} + x^{n+2+\sigma}y^{n+2+\sigma} + x^{n+3+\sigma}y^{n+3+\sigma} + \cdots$$

and so

$$\int_0^1\int_0^1 \frac{(xy)^{n+\sigma}}{1-xy}dxdy = \int_0^1\left\{\int_0^1(x^{n+\sigma}y^{n+\sigma} + x^{n+1+\sigma}y^{n+1+\sigma} + x^{n+2+\sigma}y^{n+2+\sigma} + \cdots)dx\right\}dy$$

$$= \int_0^1\left(\frac{1}{n+\sigma+1}x^{n+\sigma+1}y^{n+\sigma} + \frac{1}{n+\sigma+2}x^{n+\sigma+2}y^{n+\sigma+1}\right.$$

$$\left. + \frac{1}{n+\sigma+3}x^{n+\sigma+3}y^{n+\sigma+2} + \cdots\right)\Big|_0^1 dy$$

$$= \int_0^1\left(\frac{1}{n+\sigma+1}y^{n+\sigma} + \frac{1}{n+\sigma+2}y^{n+\sigma+1} + \frac{1}{n+\sigma+3}y^{n+\sigma+2} + \cdots\right)dy$$

$$= \left(\left\{\frac{1}{n+\sigma+1}\right\}\frac{1}{n+\sigma+1}y^{n+\sigma+1} + \left\{\frac{1}{n+\sigma+2}\right\}\frac{1}{n+\sigma+2}y^{n+\sigma+2}\right.$$

$$\left. + \left\{\frac{1}{n+\sigma+3}\right\}\frac{1}{n+\sigma+3}y^{n+\sigma+3} + \cdots\right)\Big|_0^1$$

$$= \frac{1}{(n+\sigma+1)^2} + \frac{1}{(n+\sigma+2)^2} + \frac{1}{(n+\sigma+3)^2} + \cdots.$$

Thus,

$$I(\sigma) = \sum_{k=1}^{\infty} \frac{1}{(n+\sigma+k)^2}$$

and so

$$\frac{dI}{d\sigma} = \sum_{k=1}^{\infty} \frac{-2(n+\sigma+k)}{(n+\sigma+k)^4} = -2\sum_{k=1}^{\infty} \frac{1}{(n+\sigma+k)^3},$$

which says

$$\frac{dI}{d\sigma}\Big|_{\sigma=0} = \int_0^1 \int_0^1 \frac{(xy)^n \ln(xy)}{1-xy} dx\, dy = -2\sum_{k=1}^{\infty} \frac{1}{(n+k)^3}$$

$$= -2\left[\left(\frac{1}{1^3} + \frac{1}{2^3} + \cdots + \frac{1}{n^3}\right) + \left(\frac{1}{(n+1)^3} + \frac{1}{(n+2)^3} + \cdots\right) - \left(\frac{1}{1^3} + \frac{1}{2^3} + \cdots + \frac{1}{n^3}\right)\right]$$

$$= -2\left[\zeta(3) - \left(\frac{1}{1^3} + \frac{1}{2^3} + \cdots + \frac{1}{n^3}\right)\right] = 2\left(\frac{1}{1^3} + \frac{1}{2^3} + \cdots + \frac{1}{n^3}\right) - 2\zeta(3).$$

So, for the cases of $n = 1$ and $n = 2$, for example,

$$\int_0^1 \int_0^1 \frac{xy \ln(xy)}{1-xy} dx\, dy = 2 - 2\zeta(3)$$

and

$$\int_0^1 \int_0^1 \frac{x^2 y^2 \ln(xy)}{1-xy} dx\, dy = 2\left(1 + \frac{1}{8}\right) - 2\zeta(3) = \frac{9}{4} - 2\zeta(3) = \left(\frac{3}{2}\right)^2 - 2\zeta(3).$$

In 1772 Euler came as close as he ever would to $\zeta(3)$ when he stated

$$\int_0^{\pi/2} x \ln\{\sin(x)\}dx = \frac{7}{16}\zeta(3) - \frac{\pi^2}{8}\ln(2).$$

The key to understanding how such a stunning equality could be discovered is Euler's identity. Here's how it goes. Define the function $S(y)$ as

$$S(y) = 1 + e^{iy} + e^{i2y} + e^{i3y} + \cdots + e^{imy}$$

where m is some finite integer. This looks like a geometric series and so, using the standard trick for summing such series, multiply through by the common factor e^{iy} that connects any two adjacent terms. Then,

$$e^{iy}S(y) = e^{iy} + e^{i2y} + e^{i3y} + \cdots + e^{imy} + e^{i(m+1)y}$$

and so

$$e^{iy}S(y) - S(y) = e^{i(m+1)y} - 1.$$

Solving for $S(y)$,

$$S(y) = \frac{e^{i(m+1)y} - 1}{e^{iy} - 1} = \frac{e^{i(m+1)y} - 1}{e^{i\frac{y}{2}}\left(e^{i\frac{y}{2}} - e^{-i\frac{y}{2}}\right)} = \frac{e^{i\left(m+\frac{1}{2}\right)y} - e^{-i\frac{y}{2}}}{i2\sin\left(\frac{y}{2}\right)}$$

or

$$(2.4.19) \quad S(y) = \frac{\cos\left\{\left(m+\frac{1}{2}\right)y\right\} + i\sin\left\{\left(m+\frac{1}{2}\right)y\right\} - \cos\left(\frac{y}{2}\right) + i\sin\left(\frac{y}{2}\right)}{i2\sin\left(\frac{y}{2}\right)}.$$

Now, looking back at the original definition of $S(y)$, we see that it can also be written as

$$(2.4.20) \qquad S(y) = 1 + \sum_{n=1}^{m} \cos(ny) + i \sum_{n=1}^{m} \sin(ny).$$

So, equating the imaginary parts of our two alternative expressions for $S(y)$, (2.4.19) and (2.4.20), we have

$$-\frac{\cos\left\{\left(m+\frac{1}{2}\right)y\right\}}{2\sin\left(\frac{y}{2}\right)} + \frac{\cos\left(\frac{y}{2}\right)}{2\sin\left(\frac{y}{2}\right)} = \sum_{n=1}^{m} \sin(ny).$$

At this point it is convenient to change variable to $y = 2t$, and so

$$-\frac{\cos\{(2m+1)t\}}{\sin(t)} + \cot(t) = 2 \sum_{n=1}^{m} \sin(2nt).$$

Then, integrate this expression, term-by-term, from $t = x$ to $t = \frac{\pi}{2}$, getting

$$-\int_{x}^{\frac{\pi}{2}} \frac{\cos\{(2m+1)t\}}{\sin(t)} dt + \int_{x}^{\frac{\pi}{2}} \cot(t) dt = 2 \sum_{n=1}^{m} \int_{x}^{\frac{\pi}{2}} \sin(2nt) dt.$$

The integral on the right of the equality sign is easy to do:

$$\int_{0}^{\frac{\pi}{2}} \sin(2nt) dt = \left\{ -\frac{\cos(2nt)}{2n} \right\} \Big|_{x}^{\frac{\pi}{2}}$$

$$= \frac{-\cos(n\pi) + \cos(2nx)}{2n} = \frac{\cos(2nx) - (-1)^n}{2n}.$$

The last integral on the left of the equality sign is just as easy to do:

$$\int_x^{\frac{\pi}{2}} \cot(t)dt = [\ln\{\sin(t)\}]\Big|_x^{\frac{\pi}{2}} = \ln\left\{\sin\left(\frac{\pi}{2}\right)\right\}$$

$$-\ln\{\sin(x)\} = -\ln\{\sin(x)\}.$$

Thus,

(2.4.21)

$$-\int_x^{\frac{\pi}{2}} \frac{\cos\{(2m+1)t\}}{\sin(t)} dt - \ln\{\sin(x)\}$$

$$= \sum_{n=1}^m \frac{\cos(2nx)}{n} - \sum_{n=1}^m \frac{(-1)^n}{n}.$$

Next, recall the power series expansion

$$\ln(1+x) = x - \frac{x^2}{2} + \frac{x^3}{3} - \frac{x^4}{4} + \cdots$$

and so, with $x = 1$, this says

$$\ln(2) = -\sum_{n=1}^\infty \frac{(-1)^n}{n}.$$

So, if we let $m \to \infty$ in (2.4.21), we have

$$-\lim_{m\to\infty} \int_x^{\frac{\pi}{2}} \frac{\cos\{(2m+1)t\}}{\sin(t)} dt - \ln\{\sin(x)\} = \sum_{n=1}^\infty \frac{\cos(2nx)}{n} + \ln(2).$$

Since the limit of the integral at the far left is zero,[20] we arrive at

$$(2.4.22) \qquad \ln\{\sin(x)\} = -\sum_{n=1}^{\infty} \frac{\cos(2nx)}{n} - \ln(2).$$

The next step (one not particularly obvious, but remember that we are following in *Euler*'s footsteps!) is to first multiply through (2.4.22) by x and then integrate from 0 to $\frac{\pi}{2}$. That is, let's write

$$\int_0^{\frac{\pi}{2}} x \ln\{\sin(x)\} dx = -\sum_{n=1}^{\infty} \frac{1}{n} \int_0^{\frac{\pi}{2}} x \cos(2nx) dx - \ln(2) \int_0^{\frac{\pi}{2}} x \, dx$$

$$= -\sum_{n=1}^{\infty} \frac{1}{n} \int_0^{\frac{\pi}{2}} x \cos(2nx) dx - \frac{\pi^2}{8} \ln(2).$$

To do the integral on the right, use integration by parts, with $u = x$ and $dv = \cos(2nx) \, dx$. I'll let you fill in the details to show that

$$\int_0^{\frac{\pi}{2}} x \, \cos(2nx) dx = \begin{cases} -\dfrac{1}{2n^2}, \text{ if } n \text{ is odd} \\[2mm] 0, \text{ if } n \text{ is even.} \end{cases}$$

Thus,

$$(2.4.23) \qquad \int_0^{\frac{\pi}{2}} x \ln\{\sin(x)\} dx = \frac{1}{2} \left\{ \sum_{n=1,n \text{ odd}}^{\infty} \frac{1}{n^3} \right\} - \frac{\pi^2}{8} \ln(2).$$

20. This assertion follows from the almost intuitively obvious ("obvious" if you invoke the area interpretation of the integral) *Riemann-Lebesgue lemma*—after Riemann, of course, and the French mathematician Henri Lebesgue (1875–1941)—which says that if $f(t)$ is *absolutely integrable* over the interval a to b, then $\lim_{m \to \infty} \int_a^b f(t) \cos(mt) dt = \lim_{m \to \infty} \int_a^b f(t) \sin(mt) dt = 0$. Physicists and engineers are typically okay with an "obvious" area interpretation of the lemma, but mathematicians typically (and rightfully, I have to admit) call it borderline handwaving, and they want a more substantial justification. If you feel that way, too, take a look at Georgi P. Tolstov, *Fourier Series* (Dover, 1976), pp. 70–71. We'll briefly revisit justifying the lemma in the next chapter (see Challenge Problem 3.3.3).

We are now almost done. All that's left to do is to note that

$$\sum_{n=1, n \text{ even}}^{\infty} \frac{1}{n^3} = \frac{1}{2^3} + \frac{1}{4^3} + \frac{1}{6^3} + \cdots = \frac{1}{(2 \cdot 1)^3} + \frac{1}{(2 \cdot 2)^3} + \frac{1}{(2 \cdot 3)^3} + \cdots$$

$$= \frac{1}{8}\left(\frac{1}{1^3} + \frac{1}{2^3} + \frac{1}{3^3} + \cdots \right)$$

$$= \frac{1}{8}\sum_{n=1}^{\infty} \frac{1}{n^3} = \frac{1}{8}\zeta(3).$$

And so, since

$$\sum_{n=1, n \text{ odd}}^{\infty} \frac{1}{n^3} + \sum_{n=1, n \text{ even}}^{\infty} \frac{1}{n^3} = \zeta(3),$$

we have

(2.4.24) $$\sum_{n=1, n \text{ odd}}^{\infty} \frac{1}{n^3} = \zeta(3) - \sum_{n=1, n \text{ even}}^{\infty} \frac{1}{n^3} = \zeta(3) - \frac{1}{8}\zeta(3) = \frac{7}{8}\zeta(3).$$

Thus, just as Euler declared, putting (2.4.24) into (2.4.23), we have

(2.4.25) $$\int_0^{\pi/2} x \ln\{\sin(x)\}dx = \frac{7}{16}\zeta(3) - \frac{\pi^2}{8}\ln(2).$$

Alas, nobody—not even Euler—has been able to do the integral in (2.4.25).[21]

Challenge Problem 2.4.1: Calculate the value of $\int_0^1 \int_0^1 \frac{dxdy}{1-x^2y^2} = ?$ Hint: This is easy, if you think of using P and Q together.

---

21. To add some frustration (as if we actually need more) to this point, if we increase the upper limit to π, then it is well known that $\int_0^\pi x \ln\{\sin(x)\}dx = -\frac{\pi^2}{2}\ln(2)$. Are there no limits to how far the number gods will go to taunt us?

Challenge Problem 2.4.2: Calculate the value of $\int_0^1 \int_0^1 \frac{(xy)^n}{1-xy} dxdy = ?$
Hint: For $n = 3$, your answer should reduce to $\zeta(2) - (\frac{7}{6})^2$.

Challenge Problem 2.4.3: Show that $\int_0^1 \int_0^1 \int_0^1 \frac{dxdydz}{1+xyz} = \frac{3}{4}\zeta(3)$.

2.5 Zeta *Near s* = 1

We know that $\zeta(1) = \infty$, because when $s = 1$, $\zeta(s)$ reduces to the divergent harmonic series. That is,

$$\lim_{s \to 1} \zeta(s) = \infty.$$

An interesting calculation to perform is to determine how $\zeta(s)$ blows up as s approaches 1. Since the harmonic series diverges logarithmically, does that mean $\zeta(s)$ blows up logarithmically, too, as $s \to 1$? *No*, and in fact, $\zeta(s)$ blows up faster than logarithmically as $s \to 1$. This may seem paradoxical, but by the time we reach the end of this section, it may not seem quite so puzzling.

Here's our plan of attack. From (2.1.9) and (2.1.10) we can write

$$\eta(s) = \sum_{k=1}^{\infty} \frac{(-1)^{k+1}}{k^s} = (1 - 2^{1-s})\zeta(s)$$

and so

(2.5.1) $$\zeta(s) = \frac{1}{(1 - 2^{1-s})} \sum_{k=1}^{\infty} \frac{(-1)^{k+1}}{k^s}.$$

We'll then write $s = 1 + \varepsilon$ and, finally, we'll let $\varepsilon \to 0$ and see what happens. Since $1 - s = -\varepsilon$, then

$$\frac{1}{(1 - 2^{1-s})} = \frac{1}{(1 - 2^{-\varepsilon})} = \frac{1}{1 - e^{\ln(2^{-\varepsilon})}} = \frac{1}{1 - e^{-\varepsilon \ln(2)}}.$$

From the power series expansion of the exponential, we have

$$1 - e^{-\varepsilon \ln(2)} = 1 - \left\{ 1 - \varepsilon \ln(2) + \frac{\varepsilon^2 \ln^2(2)}{2!} - \cdots \right\}$$

$$= \varepsilon \ln(2) - \frac{1}{2} \varepsilon^2 \ln^2(2)$$

± terms in higher powers of ε,

and as $\varepsilon \to 0$ these additional terms become negligible compared to the first two terms. So, with decreasing error as $\varepsilon \to 0$, we have

$$1 - e^{-\varepsilon \ln(2)} \approx \varepsilon \ln(2) \left[1 - \frac{1}{2} \varepsilon \ln(2) \right].$$

Thus, as $\varepsilon \to 0$ (as $s \to 1$),

$$\frac{1}{(1 - 2^{1-s})} \approx \frac{1}{\varepsilon \ln(2) \left[1 - \frac{1}{2} \varepsilon \ln(2) \right]}$$

or

(2.5.2) $$\frac{1}{(1 - 2^{1-s})} \approx \frac{1}{\varepsilon \ln(2)} \left[1 + \frac{1}{2} \varepsilon \ln(2) \right].$$

Next, let's see what happens in the summation of (2.5.1) as $s \to 1$. We have

$$\frac{1}{k^s} = \frac{1}{k^{1+\varepsilon}} = \frac{1}{k k^\varepsilon} = \frac{1}{k e^{\ln(k^\varepsilon)}} = \frac{1}{k e^{\varepsilon \ln(k)}} = \frac{1}{k \{ 1 + \varepsilon \ln(k) + \cdots \}}$$

or, as $\varepsilon \to 0$ ($s \to 1$), we see that

$$\frac{1}{k^s} \approx \frac{1}{k}[1 - \varepsilon \ln(k)].$$

Thus,

$$\sum_{k=1}^{\infty} \frac{(-1)^{k+1}}{k^s} \approx \sum_{k=1}^{\infty} \frac{(-1)^{k+1}}{k}[1 - \varepsilon \ln(k)]$$

$$= \sum_{k=1}^{\infty} \frac{(-1)^{k+1}}{k} - \varepsilon \sum_{k=1}^{\infty} \frac{(-1)^{k+1}}{k} \ln(k).$$

Recognizing the first sum on the right as $\ln(2)$—just plug $x = 1$ into the power series expansion of $\ln(1 + x)$—we have, as $\varepsilon \to 0$ $(s \to 1)$, that

(2.5.3) $$\sum_{k=1}^{\infty} \frac{(-1)^{k+1}}{k^{1+\varepsilon}} \approx \ln(2) - \varepsilon \sum_{k=1}^{\infty} \frac{(-1)^{k+1}}{k} \ln(k).$$

If we plug (2.5.2) and (2.5.3) into (2.5.1), we see that

$$\zeta(1+\varepsilon) \approx \frac{1}{\varepsilon \ln(2)}\left[1 + \frac{1}{2}\varepsilon \ln(2)\right]\left[\ln(2) - \varepsilon \sum_{k=1}^{\infty} \frac{(-1)^{k+1}}{k} \ln(k)\right]$$

$$= \left[\frac{1}{\varepsilon \ln(2)} + \frac{1}{2}\right]\left[\ln(2) - \varepsilon \sum_{k=1}^{\infty} \frac{(-1)^{k+1}}{k} \ln(k)\right]$$

$$= \frac{1}{\varepsilon} - \frac{1}{\ln(2)} \sum_{k=1}^{\infty} \frac{(-1)^{k+1}}{k} \ln(k) + \frac{1}{2}\ln(2)$$

+ terms of all powers of ε, from the first.

Thus, as $\varepsilon \to 0$,

(2.5.4) $$\zeta(1+\varepsilon) \approx \frac{1}{\varepsilon} + \frac{1}{2}\ln(2) - \frac{1}{\ln(2)} \sum_{k=1}^{\infty} \frac{(-1)^{k+1}}{k} \ln(k).$$

Returning to $s = 1 + \varepsilon$ and writing $(-1)^{k+1} = -(-1)^k$, (2.5.4) becomes (as $s \to 1$)

$$\zeta(s) \approx \frac{1}{s-1} + \frac{1}{2}\ln(2) + \frac{1}{\ln(2)}\sum_{k=1}^{\infty}\frac{(-1)^k}{k}\ln(k)$$

and so, as $\ln(1) = 0$, then as $s \to 1$, we have

(2.5.5) $$\zeta(s) \approx \frac{1}{s-1} + \frac{1}{2}\ln(2) + \frac{1}{\ln(2)}\sum_{k=2}^{\infty}\frac{(-1)^k}{k}\ln(k).$$

Whatever the value of the summation term at the far right of (2.5.5) may be, we know it is finite because of note 6 in Chapter 1 (because $\ln(k)$ grows more slowly than does k). That is, as $s \to 1$,

$$\zeta(s) \approx \frac{1}{s-1} + \text{constant},$$

where the constant is given by

(2.5.6) $$\text{constant} = \frac{1}{2}\ln(2) + \frac{1}{\ln(2)}\sum_{k=2}^{\infty}\frac{(-1)^k}{k}\ln(k).$$

If we numerically evaluate (2.5.6), using the first 1 million terms of the sum, we get $0.5772\ldots$, and it is very hard to resist the temptation to suspect that the constant is actually Euler's constant, γ. Here's how to prove that is, indeed, the case.

   We start with the calculation of a simple integral: for n any positive integer,

$$\int_n^{\infty}\frac{1}{x^{s+1}}\,dx = \int_n^{\infty}x^{-s-1}\,dx = -\frac{x^{-s}}{s}\Big|_n^{\infty} = \frac{1}{sn^s}$$

and so

$$\frac{1}{n^s} = s\int_n^\infty \frac{1}{x^{s+1}}dx.$$

Thus,

$$\zeta(s) = \sum_{n=1}^\infty \frac{1}{n^s} = \sum_{n=1}^\infty s\int_n^\infty \frac{1}{x^{s+1}}dx.$$

That is, if we write

$$\int_n^\infty \frac{1}{x^{s+1}}dx = \lim_{k\to\infty}\int_n^k \frac{1}{x^{s+1}}dx,$$

then

$$\zeta(s) = \sum_{n=1}^\infty s\lim_{k\to\infty}\int_n^k \frac{1}{x^{s+1}}dx$$

or, since n doesn't exceed k in the integral, we can stop the sum at $n = k$ and so we arrive at

(2.5.7) $$\zeta(s) = s\lim_{k\to\infty}\sum_{n=1}^k \int_n^k \frac{1}{x^{s+1}}dx.$$

Next, write out (2.5.7) term-by-term, in the form of a matrix of $k-1$ rows, with each row corresponding to a value of n as n varies from 1 (the top row) to $k-1$ (the bottom row; as when $n = k$, the associated integral is zero and so can be ignored):

$$\sum_{n=1}^k \int_n^k \frac{1}{x^{s+1}}dx = \begin{bmatrix} \int_1^2 & +\int_2^3 + & \int_3^4 + & \dots & +\int_{k-1}^k \\ & +\int_2^3 + & \int_3^4 + & \dots & +\int_{k-1}^k \\ & & +\int_3^4 + & \dots & +\int_{k-1}^k \\ & & & \vdots & \\ & & & & +\int_{k-1}^k \end{bmatrix}.$$

Now, add all these terms vertically (that is, in columns, going left to right), to get

$$\sum_{n=1}^{k} \int_{n}^{k} \frac{1}{x^{s+1}} dx = \int_{1}^{2} \frac{1}{x^{s+1}} dx + 2\int_{2}^{3} \frac{1}{x^{s+1}} dx + 3\int_{3}^{4} \frac{1}{x^{s+1}} dx + \cdots$$

$$+ (k-1)\int_{k-1}^{k} \frac{1}{x^{s+1}} dx$$

$$= \int_{1}^{2} \frac{1}{x^{s+1}} dx + \int_{2}^{3} \frac{2}{x^{s+1}} dx + \int_{3}^{4} \frac{3}{x^{s+1}} dx + \cdots + \int_{k-1}^{k} \frac{k-1}{x^{s+1}} dx.$$

Notice that the numerator of each integral equals the lower limit of that integral. That is, over the integration interval for each integral, the numerator is $\lfloor x \rfloor = x - \{x\}$ (you'll recall we discussed this notation back in Section 1.6). Thus,

$$\sum_{n=1}^{k} \int_{n}^{k} \frac{1}{x^{s+1}} dx = \int_{1}^{2} \frac{\lfloor x \rfloor}{x^{s+1}} dx + \int_{2}^{3} \frac{\lfloor x \rfloor}{x^{s+1}} dx +$$

$$\int_{3}^{4} \frac{\lfloor x \rfloor}{x^{s+1}} dx + \cdots + \int_{k-1}^{k} \frac{\lfloor x \rfloor}{x^{s+1}} dx$$

$$= \int_{1}^{k} \frac{\lfloor x \rfloor}{x^{s+1}} dx = \int_{1}^{k} \frac{x - \{x\}}{x^{s+1}} dx$$

and so, plugging this into (2.5.7) and letting $k \to \infty$,

$$\zeta(s) = s\int_{1}^{\infty} \frac{x - \{x\}}{x^{s+1}} dx = s\int_{1}^{\infty} \frac{dx}{x^{s}} - s\int_{1}^{\infty} \frac{\{x\}}{x^{s+1}} dx.$$

Since

$$\int_{1}^{\infty} \frac{dx}{x^{s}} = \int_{1}^{\infty} x^{-s} dx = \left(\frac{x^{-s+1}}{-s+1}\right)\Big|_{1}^{\infty} = -\frac{1}{1-s} = \frac{1}{s-1},$$

then

$$(2.5.8) \qquad \zeta(s) = \frac{s}{s-1} - s\int_1^\infty \frac{\{x\}}{x^{s+1}} dx.$$

All that remains to do is the evaluation of the integral in (2.5.8) for the case of $s = 1$.

That is, let's now calculate

$$\int_1^\infty \frac{\{x\}}{x^2} dx = ?$$

The following line of mathematics comes from recognizing that, over an integration interval from one integer to the next, the fractional value of x (given by $\{x\}$) in that interval is given by x minus the lower limit. So,

$$\int_1^\infty \frac{\{x\}}{x^2} dx = \lim_{k\to\infty}\left[\int_1^2 \frac{x-1}{x^2} dx + \int_2^3 \frac{x-2}{x^2} dx + \int_3^4 \frac{x-3}{x^2} dx + \cdots + \right.$$

$$\left. \int_{k-1}^k \frac{x-(k-1)}{x^2} dx\right]$$

$$= \lim_{k\to\infty}\int_1^k \frac{dx}{x} - \lim_{k\to\infty}\left[\int_1^2 \frac{dx}{x^2} + 2\int_2^3 \frac{dx}{x^2} + 3\int_3^4 \frac{dx}{x^2} + \cdots + (k-1)\int_{k-1}^k \frac{dx}{x^2}\right]$$

$$= \lim_{k\to\infty}\left[\ln(x)|_1^k - \left\{\left(-\frac{1}{x}\right)\Big|_1^2 + 2\left(-\frac{1}{x}\right)\Big|_2^3 + 3\left(-\frac{1}{x}\right)\Big|_3^4 + \cdots + (k-1)\left(-\frac{1}{x}\right)\Big|_{k-1}^k\right\}\right]$$

$$= \lim_{k\to\infty}\left[\ln(k) - \left\{\left(1-\frac{1}{2}\right) + 2\left(\frac{1}{2}-\frac{1}{3}\right) + 3\left(\frac{1}{3}-\frac{1}{4}\right) + \cdots + (k-1)\left(\frac{1}{k-1}-\frac{1}{k}\right)\right\}\right]$$

$$= \lim_{k\to\infty}\left[\ln(k) - \left\{\frac{2-1}{2} + 2\frac{3-2}{6} + 3\frac{4-3}{12} + \cdots + (k-1)\frac{k-(k-1)}{(k-1)k}\right\}\right]$$

$$= \lim_{k \to \infty} \left[\ln(k) - \left\{ \frac{1}{2} + \frac{1}{3} + \frac{1}{4} + \cdots + \frac{1}{k} \right\} \right]$$

$$= \lim_{k \to \infty} \left[\ln(k) - \left\{ 1 + \frac{1}{2} + \frac{1}{3} + \frac{1}{4} + \cdots + \frac{1}{k} \right\} + 1 \right]$$

$$= \lim_{k \to \infty} [\ln(k) - h(k) + 1],$$

where $h(k)$ is the partial sum of the first k terms of the harmonic series. But since

$$\gamma = \lim_{k \to \infty} [h(k) - \ln(k)],$$

then just like that, we have the value of our integral:

(2.5.9) $$\int_1^\infty \frac{\{x\}}{x^2} dx = 1 - \gamma.$$

Putting (2.5.9) into (2.5.8) we have, as $s \to 1$,

$$\zeta(s) = \frac{s}{s-1} - s(1-\gamma) = \frac{s}{s-1} - s + s\gamma = \frac{s - s^2 + s}{s-1} + s\gamma = \frac{2s - s^2}{s-1} + s\gamma$$

or, as $s \to 1$,

$$\zeta(s) = \frac{1}{s-1} + \gamma,$$

which, comparing to (2.5.5), says

$$\gamma = \frac{1}{2}\ln(2) + \frac{1}{\ln(2)} \sum_{k=2}^\infty \frac{(-1)^k}{k} \ln(k).$$

Turning this around, we have the following interesting summation formula:

$$(2.5.10) \qquad \sum_{k=2}^{\infty} \frac{(-1)^k}{k} \ln(k) = \ln(2)\left\{ \gamma - \frac{1}{2}\ln(2) \right\}.$$

Challenge Problem 2.5.1: As $s \to 1$, (2.5.5) says that the zeta function blows up as $\frac{1}{s-1}$, that is, hyperbolically and not logarithmically, even though as $s \to 1$, $\zeta(s)$ becomes the logarithmically divergent harmonic series. Explain why this is not a conflict.

2.6 Zeta Prime at $s = 0$

In Challenge Problem 2.1.3, you were asked to calculate the value of $\zeta(0)$. The solution at the back of the book shows it is not really a difficult calculation (and if you haven't done it yet, *stop right now and do it*, because—spoiler alert!—I'm going to use that value in this section). After calculating $\zeta(0)$, any curious analyst would then immediately wonder about the value of the first derivative of $\zeta(s)$ at $s = 0$. That is, what is $\zeta'(0) = ?$ In fact, mathematicians have long wondered about the higher-order derivatives of $\zeta(s)$, at $s = 0$, that are far beyond merely the first.[22] The derivatives are, after all, central to deriving the power series expansion of $\zeta(s)$ around $s = 0$ (the Taylor series, about which I'll say more in Chapter 3).

To start our calculation of $\zeta'(0)$, we write, from (2.1.9) and (2.1.10),

$$(2.6.1) \qquad \eta(s) = \sum_{k=1}^{\infty} \frac{(-1)^{k+1}}{k^s} = (1 - 2^{1-s})\zeta(s).$$

(As the solution for $\zeta(0)$ in the back of the book shows, $\eta(0) = \frac{1}{2}$ and $\zeta(0) = -\frac{1}{2}$.) Taking the logarithm of (2.6.1), we have

---

22. For how to find the first 18 (!) derivatives, see Tom M. Apostol, "Formulas for Higher Derivatives of the Riemann Zeta Function," *Mathematics of Computation*, January 1985, pp. 223–232. (This is the same Apostol mentioned in Section 2.4, concerning the calculation of $\zeta(2)$.) Apostol's method for calculating $\zeta'(0)$, and then the next 17 derivatives, is far more sophisticated than what I'm showing you here.

$$\ln\{\eta(s)\} = \ln\{(1 - 2^{1-s})\} + \ln\{\zeta(s)\},$$

and so, differentiating with respect to s,

(2.6.2) $$\frac{\eta'(s)}{\eta(s)} = \frac{(1-2^{1-s})'}{1-2^{1-s}} + \frac{\zeta'(s)}{\zeta(s)}.$$

Since

$$2^{1-s} = e^{\ln(2^{1-s})} = e^{(1-s)\ln(2)} = e^{\ln(2)}e^{-s\ln(2)} = 2e^{-s\ln(2)},$$

we have

$$(1-2^{1-s})' = -2\{-\ln(2)\}e^{-s\ln(2)} = 2\ln(2)2^{-s} = 2^{1-s}\ln(2)$$

and so (2.6.2) becomes

(2.6.3) $$\frac{\zeta'(s)}{\zeta(s)} = \frac{\eta'(s)}{\eta(s)} - \frac{2^{1-s}\ln(2)}{1-2^{1-s}}.$$

Also,

$$\eta'(s) = \left\{ \sum_{k=1}^{\infty} \frac{(-1)^{k+1}}{k^s} \right\}' = \sum_{k=1}^{\infty} (-1)^{k+1} \left(\frac{1}{k^s} \right)'.$$

Since

$$\frac{1}{k^s} = k^{-s} = e^{\ln(k^{-s})} = e^{-s\ln(k)}$$

then

$$\left(\frac{1}{k^s} \right)' = -\ln(k)e^{-s\ln(k)} = -\frac{\ln(k)}{k^s}$$

128

and so

$$\eta'(s) = -\sum_{k=1}^{\infty} (-1)^{k+1} \frac{\ln(k)}{k^s}.$$

Thus,

$$\eta'(0) = -\sum_{k=1}^{\infty} (-1)^{k+1} \ln(k) = \sum_{k=1}^{\infty} (-1)^k \ln(k)$$

or, since $\ln(1) = 0$, we can drop the $k = 1$ term and write

$$\eta'(0) = \sum_{k=2}^{\infty} (-1)^k \ln(k) = \ln(2) - \ln(3) + \ln(4) - \ln(5) + \ln(6) - \cdots$$

$$= \ln\left(\frac{2 \cdot 4 \cdot 6 \cdot 8 \cdots}{3 \cdot 5 \cdot 7 \cdots}\right) = \ln\left(\sqrt{\frac{2 \cdot 2 \cdot 4 \cdot 4 \cdot 6 \cdot 6 \cdot 8 \cdot 8 \cdots}{3 \cdot 3 \cdot 5 \cdot 5 \cdot 7 \cdot 7 \cdots}}\right),$$

which, if you remember Wallis' product from Section 1.4, says

$$\eta'(0) = \ln\left(\sqrt{\frac{\pi}{2}}\right).$$

Now, since $\zeta(0) = -\frac{1}{2}$ and $\eta(0) = \frac{1}{2}$, then (2.6.3) becomes, at $s = 0$,

$$\frac{\zeta'(0)}{-\dfrac{1}{2}} = \frac{\ln\left(\sqrt{\dfrac{\pi}{2}}\right)}{\dfrac{1}{2}} - \frac{2\ln(2)}{1 - 2}$$

or

$$\zeta'(0) = -\ln\left(\sqrt{\frac{\pi}{2}}\right) - \ln(2) = -\ln\left(2\sqrt{\frac{\pi}{2}}\right) = -\ln(\sqrt{2\pi})$$

and so, finally, we have our answer:

$$(2.6.4) \qquad \zeta'(0) = -\frac{1}{2}\ln(2\pi) = -0.9189385332\ldots.$$

As a final comment on the derivatives of $\zeta(s)$ at $s = 0$, Ramanujan was the first to calculate $\zeta''(0) = \zeta^{(2)}(0)$ (in one of his private notebooks written sometime after 1900), while Apostol's results for $\zeta^{(n)}(0)$ for $n \geq 3$ were new. One interesting result in Apostol's paper is that $\zeta^{(n)}(0)/n!$ approaches -1 as $n \to \infty$, and that the convergence is pretty rapid (for example, the fourth derivative of $\zeta(s)$ at $s = 0$ is $\zeta^{(4)}(0) = -23.9971\ldots$, while $-4! = -24$).

Challenge Problem 2.6.1: In this section we made a lot of use of Dirichlet's eta function, $\eta(s)$, defined in (2.1.9). In (1.4.24) we derived Riemann's famous integral formula involving the gamma function and the zeta function: $\int_0^\infty \frac{x^{s-1}}{e^x - 1} dx = \Gamma(s)\zeta(s)$. Show that $\eta(s)$ appears in a similar integral formula: $\int_0^\infty \frac{x^{s-1}}{e^x + 1} dx = \Gamma(s)\eta(s)$.

Challenge Problem 2.6.2: You'll recall from the second box in Section 1.2 that the Riemann hypothesis says $\zeta(s) = 0$ if $s = \frac{1}{2} + ib$, where $b > 0$ is any of an infinity of particular values. Show that this says, for each such value of b, $\sum_{k=1}^{\infty} \frac{(-1)^{k+1}}{\sqrt{k}}\cos\{b\ln(k)\} = 0$ and $\sum_{k=1}^{\infty} \frac{(-1)^{k+1}}{\sqrt{k}}\sin\{b\ln(k)\} = 0$. For example, the first value of b (that is, the imaginary part of the first complex zero of the zeta function) is $b = 14.13472514173469379\ldots$), and summing the first 100 million terms of each sum, for that value of b, gives, respectively, 4.64×10^{-5} and -1.86×10^{-5}. Not *exactly* zero, but pretty small (the $\sqrt{k}$ in the denominators slows the convergence down). Hint: Use (2.6.1) and Euler's identity.

Challenge Problem 2.6.3: As discussed in the second box of Section 1.2, Hardy showed in 1914 that the zeta function has an infinite number of zeros on the critical line. Show, in contrast, that the gamma function vanishes nowhere on the critical line. Hint: The reflection formula for the gamma function is a good place to start, and Euler's identity will prove useful, too.

2.7 Interlude

Before we continue to the next chapter and plunge ever more deeply into the zeta function, let's take a brief break. In this final section of what I've called a "wizard math" chapter, I'll revisit the issue of just why it is the calculation of zeta-3 that is the world's "most puzzling" math problem, over all other possibilities. When I started writing, I must admit there was in my mind one other possible candidate problem that easily met three of my four selection criteria (I'll say more, soon, about the fourth criterion it doesn't satisfy): it certainly is an unsolved problem, people have been trying to solve it for centuries, and a grammar school student can easily understand it. And, as an added treat, Euler is again involved!

I'm speaking of the question of the infinity (or not) of *perfect numbers*. A positive integer is called *perfect* if it is equal to the sum of its positive divisors, starting with 1 but *not* including itself. That's it, and any youngster who is able to divide and add is technically equipped with everything she needs to attack this problem. Indeed, the first four perfect numbers were known long ago to the ancient (circa A.D. 100) Greeks:

$$6 = 1 + 2 + 3; 28 = 1 + 2 + 4 + 7 + 14;$$

$$496 = 1 + 2 + 4 + 8 + 16 + 31 + 62 + 124 + 248;$$

$$8,128 = 1 + 2 + 4 + 8 + 16 + 32 + 64 + 127 + 254 + 508 + 1,016 + 2,032 + 4,064.$$

The next (fifth) perfect number wasn't discovered until many centuries later, in 1461 (by an unknown scholar): $2^{12}(2^{13} - 1) = 33,550,336$. It then took well over a century more, to 1603, for the Italian Pietro Cataldi (1548–1626) to find the sixth and seventh perfect numbers:

$$2^{16}(2^{17} - 1) = 8,589,869,056$$

and

$$2^{18}(2^{19} - 1) = 137{,}438{,}691{,}328.$$

You'll notice two things about how I have written these last three perfect numbers. First, they are of the form $2^{p-1}(2^p - 1)$, where both p and $2^p - 1$ are prime (in this case, $2^p - 1$ is called a *Mersenne prime* after the French monk Father Marin Mersenne (1588–1648)).[23] The fact that any number with this form is perfect had been proven by Euclid in his *Elements* (4th century B.C.).[24] That is, Euclid showed that a form of $2^{p-1}(2^p - 1)$ with both p and $2^p - 1$ prime is sufficient to ensure the associated even number is perfect. It left open the question of necessity: could there be even perfect number(s) with a different form? The answer is *no*, because Euler proved (in a paper published decades after his death) that all even perfect numbers have this form.[25] And that's why, for my second point, I haven't bothered to list all the divisors of these large perfect numbers, as there is no need to check their perfection by adding the divisors, because Euler proved they *must* add correctly. Euler himself found the eighth perfect number in 1772: $2^{30}(2^{31} - 1) = 2{,}305{,}843{,}008{,}139{,}952{,}128$.

As you can see from just the first eight perfect numbers, they get very big, pretty fast. This quickly motivates the question: Are there an infinite number of perfect numbers? Nobody knows, and as I write (March 2021), powerful supercomputer searches have extended our knowledge of particular perfect numbers to a total of just 51. The largest is, as will be no surprise, *really big* (using the prime $p = 82{,}589{,}933$, $2^{p-1}(2^p - 1)$ generates a perfect number with 49,724,095 digits!) Interestingly, all 51 known perfect numbers

23. Not all numbers of the form $2^p - 1$ are prime if p is prime. Many early mathematicians erroneously believed that, but in 1536, the Greek mathematician Hudalrichus Regius gave the counterexample $2^{11} - 1 = 2{,}047 = (23)(89)$.

24. You can find Euclid's proof (it's strictly high school algebra) in C. S. Ogilvy and J. T. Anderson, *Excursions in Number Theory* (Dover, 1988), pp. 21–22.

25. For the first four perfect numbers, notice that using the primes $p = 2, 3, 5$, and 7 gives $6 = 2(2^2 - 1)$, $28 = 2^2(2^3 - 1)$, $496 = 2^4(2^5 - 1)$, and $8{,}128 = 2^6(2^7 - 1)$. You can find Euler's proof in William Dunham, *Euler, The Master of Us All* (Mathematical Association of America, 1999), pp. 10–11.

are even, and so another unanswered question asks: Is there at least one *odd* perfect number? In 2012 it was shown that if there is an odd perfect number, then it must be larger than 10^{1500} and have at least 101 prime factors (not necessarily distinct). I therefore think it safe to say that nobody is likely to discover an odd perfect number (if it exists) by random doodling! The math historian John Stillwell has written (in his 2010 book *Mathematics and Its History*, Springer, p. 40) that these "may be the oldest open problem[s] in mathematics." They certainly sound like problems that should, you'd think, give the zeta-3 problem a good challenge for being declared the most puzzling. So why did I pass over perfect numbers in favor of $\zeta(3)$?

It's a judgment call, I'll admit, but I'll let the English mathematician Peter Barlow (1776–1862) answer for me. When he wrote of Euler's eighth perfect number in his 1811 book *Theory of Numbers*, he declared "It is the greatest that will ever be discovered, for, *as they are merely curious without being useful* [my emphasis], it is not likely that any person will attempt to find one beyond it."[26] In that judgment he was clearly in error, not being able to anticipate the irresistible urge to employ the fantastic computational power of the electronic computers that would appear less than 140 years later. But his central point is still valid. With the discovery of each new, larger perfect number, all we have is a new, bigger number with no apparent role to play in anything to do with science or engineering. Purists may rejoice—and I am certainly not one to sneer—but compare that situation to the muscular role of the zeta function in both mathematics and science (particularly physics).

---

26. Barlow's book was published by J. Johnson of London (the quote is from p. 43). This rarity of the perfect numbers prompted the religious Mersenne to declare (in 1644, when only seven perfect numbers were known) "We see clearly ... how rare are the Perfect Numbers and how right we are to compare them with perfect men." Modern mathematicians, even if religious, are less inclined to attach such metaphysical commentary to their discoveries.

One other famous problem that satisfies all four of my selection criteria, but which I nonetheless decided to also be unsuitable for this book (because, in a way, it has been solved—sort of) is the *four-color planar map problem*. First posed in 1852 by the English mathematician Augustus De Morgan (1806–1871), the problem asks for a proof of the theorem that four colors are necessary and sufficient to color all possible planar maps with the assurance that if two countries share a border, they can have different colors. The theorem says, in other words, that there are planar maps for which three colors are not enough, and that it is impossible to draw a planar map that requires more than four colors. When told of this claim, even people who have no interest in mathematics can't resist the urge to start sketching complicated maps to see whether they can draw one that needs more than four colors. (Do I see *you* looking for a pencil, a sheet of paper, and a box of crayons, right now?!) A proof remained elusive for a long time, even though the apparently more difficult three-dimensional version for maps drawn on a torus (a donut surface) has been solved (seven colors are necessary and sufficient). Then, in 1976, two mathematicians at the University of Illinois programmed a computer to systematically check all of the many hundreds of specific special cases to which they had reduced all possible planar maps (K. Appel and W. Haken, "The Four Color Proof Suffices," *Mathematical Intelligencer* 1986 (no. 1), pp. 10–20). Most mathematicians today, even if they accept the conclusion that the theorem is correct, consider this approach by computer to be a concession to barbarism. The two programmers themselves told the story of the horrified reaction of a mathematician friend who exclaimed, when informed about how they had solved the problem, "God would never permit the best proof of such a beautiful theorem to be so ugly!" So, for most mathematicians the four-color planar map theorem still waits to be given a proper proof. But even if so solved, the planar map theorem appears to lack (as do perfect

numbers) the applicability that the zeta function has to engineering and science. (Ken Appel (1932–2013) was chairman of the
math department at the University of New Hampshire in the
1990s, and I know he was quite proud of his computer proof.)

Challenge Problem 2.7.1: To keep my claim valid that every section in this book has at least one challenge problem, here are three
simple (but quite interesting) facts for you to prove concerning perfect numbers. (1) Suppose p and q are any two distinct, odd primes.
That is, $p \neq q$, and neither is equal to 2. Prove that the product pq
can*not* be perfect. Hint: First, notice that the only divisors of pq are
$1, p$, and q, and then convince yourself that $(p - 1)(q - 1) > 2$. You're
now just one step away from your goal. (2) Also, prove that no integer power k of any prime p can be a perfect number. Hint: With
$n = p^k$, $k \geq 1$, the divisors of n are $1, p, p^2, p^3, \ldots, p^{k-1}$. (3) Finally,
explain why no perfect square can be an even perfect number. Hint:
Apply the Euclid-Euler result, that all even perfect numbers have the
form $2^{p-1}(2^p - 1)$, where p is prime and so is $2^p - 1$.

ANALYSIS OF CHALLENGE PROBLEM 1.8.3

The answer is $\sum_{n=1}^{\infty} \frac{1}{n(n+1)(n+2)\cdots(n+p)} = \frac{1}{p} \cdot \frac{1}{p!}$, for p any positive integer. As
the starting point for deriving this result, we return to the beta func-
tion integral discussed in (2.2.1) through (2.2.5):

$$\int_0^1 x^{m-1}(1-x)^{n-1}dx = \frac{\Gamma(m)\Gamma(n)}{\Gamma(m+n)}.$$

From this we can write, recalling (1.4.4) on how the gamma function
works with positive integer arguments,

$$\int_0^1 x^{m-1}(1-x)^n dx = \frac{\Gamma(m)\Gamma(n+1)}{\Gamma(m+n+1)} = \frac{(m-1)!n!}{(m+n)!}$$

$$= \frac{m!n!}{m(m+n)!} = \frac{n!}{m(m+1)(m+2)\cdots(m+n)}.$$

However, from the binomial theorem, we have

$$\int_0^1 x^{m-1}(1-x)^n \, dx = \int_0^1 x^{m-1} \sum_{k=0}^n \binom{n}{k}(-x)^k (1)^{n-k} \, dx$$

$$= \int_0^1 \sum_{k=0}^n \binom{n}{k}(-1)^k x^{m+k-1} dx = \sum_{k=0}^n \binom{n}{k}(-1)^k \int_0^1 x^{m+k-1} dx$$

$$= \sum_{k=0}^n \binom{n}{k}(-1)^k \left(\frac{x^{m+k}}{m+k} \right) \Big|_0^1 = \sum_{k=0}^n \binom{n}{k}(-1)^k \frac{1}{m+k}.$$

Thus, equating our two expressions for $\int_0^1 x^{m-1}(1-x)^n \, dx$, we have

$$\sum_{k=0}^n \binom{n}{k}(-1)^k \frac{1}{m+k} = \frac{n!}{m(m+1)(m+2)\cdots(m+n)}.$$

Or, if we change notation (to agree with that of Challenge Problem 1.8.3) by replacing m with n, and replacing each *original n* with a p, we have

$$\sum_{k=0}^p \binom{p}{k}(-1)^k \frac{1}{n+k} = \frac{p!}{n(n+1)(n+2)\cdots(n+p)}$$

and so

$$\frac{1}{p!} \sum_{k=0}^p \binom{p}{k}(-1)^k \frac{1}{n+k} = \frac{1}{n(n+1)(n+2)\cdots(n+p)}.$$

Then, summing over all n from 1 to infinity, we have

(1) $\quad \sum_{n=1}^\infty \frac{1}{n(n+1)(n+2)\cdots(n+p)} = \frac{1}{p!} \sum_{n=1}^\infty \sum_{k=0}^p \binom{p}{k}(-1)^k \frac{1}{n+k}.$

Now, recall the binomial coefficient identity[27]

$$\binom{p}{k} = \binom{p-1}{k} + \binom{p-1}{k-1},$$

from which it immediately follows that

$$(-1)^k \binom{p}{k} = (-1)^k \binom{p-1}{k} + (-1)^k \binom{p-1}{k-1} = (-1)^k \binom{p-1}{k} - (-1)^{k-1} \binom{p-1}{k-1}.$$

Thus, the right-most sum in (1) is (2)

$$\sum_{k=0}^{p} \binom{p}{k}(-1)^k \frac{1}{n+k} = \sum_{k=0}^{p} \binom{p-1}{k}(-1)^k \frac{1}{n+k} - \sum_{k=0}^{p} \binom{p-1}{k-1}(-1)^{k-1} \frac{1}{n+k}.$$

In the right-most sum of (2), change the index to $j = k - 1$, so that the sum becomes

$$(3) \quad \sum_{j=-1}^{p-1} \binom{p-1}{j}(-1)^j \frac{1}{n+j+1} = \sum_{j=0}^{p-1} \binom{p-1}{j}(-1)^j \frac{1}{n+j+1}$$

where the initial value for j has been increased from -1 to 0 because $\binom{p-1}{-1} = 0$ (think, *physically*, about how many ways you can select -1 books from $p - 1$ books). For the first sum on the right of (2), we write

---

27. This identity can be easily confirmed by expanding the binomial coefficients and simplifying, but it has a simple *physical* interpretation, as well, that explains how one might have originally even thought of writing such an equality. Recall that $\binom{p}{k}$ is the number of different ways to select k objects from p distinct objects $(k \leq p)$ when the order of selection is unimportant. Imagine the objects as p different books, around one of which you have tied a yellow ribbon. After picking k books, the one with the ribbon is either not one of the selected books, or it is. If it's not, then all the selected books came from the $p - 1$ books that don't have the ribbon, for a total of $\binom{p-1}{k}$ ways. If, however, it *is* among the selected books, then the remaining $k - 1$ selected books came from the $p - 1$ books that don't have the ribbon, for a total of $\binom{p-1}{k-1}$ ways. Since these two possibilities are both inclusive and mutually exclusive, the identity immediately follows.

(4) $\qquad \sum_{k=0}^{p} \binom{p-1}{k}(-1)^k \frac{1}{n+k} = \sum_{k=0}^{p-1} \binom{p-1}{k}(-1)^k \frac{1}{n+k}$

where the final value for k has been reduced from p to $p-1$, because $\binom{p-1}{p}=0$ (again think, physically, about how many ways you can select p books from $p-1$ books). Thus, using (3) and (4), (2) becomes

$$\sum_{k=0}^{p} \binom{p}{k}(-1)^k \frac{1}{n+k} = \sum_{k=0}^{p-1} \binom{p-1}{k}(-1)^k \frac{1}{n+k}$$

$$-\sum_{j=0}^{p-1} \binom{p-1}{j}(-1)^j \frac{1}{n+j+1}$$

or

(5) $\qquad \sum_{k=0}^{p} \binom{p}{k}(-1)^k \frac{1}{n+k} = \sum_{k=0}^{p-1} \binom{p-1}{k}(-1)^k \left\{ \frac{1}{n+k} - \frac{1}{n+k+1} \right\}.$

With this, (1) becomes

$$\sum_{n=1}^{\infty} \frac{1}{n(n+1)(n+2)\cdots(n+p)}$$

$$= \frac{1}{p!} \lim_{s\to\infty} \sum_{n=1}^{s} \sum_{k=0}^{p-1} \binom{p-1}{k}(-1)^k \left\{ \frac{1}{n+k} - \frac{1}{n+k+1} \right\}$$

$$= \frac{1}{p!} \sum_{k=0}^{p-1} \binom{p-1}{k}(-1)^k \lim_{s\to\infty} \sum_{n=1}^{s} \left\{ \frac{1}{n+k} - \frac{1}{n+k+1} \right\}.$$

But, as

$$\sum_{n=1}^{s}\left\{\frac{1}{n+k}-\frac{1}{n+k+1}\right\}$$

$$=\left(\frac{1}{1+k}-\frac{1}{2+k}\right)+\left(\frac{1}{2+k}-\frac{1}{3+k}\right)+\left(\frac{1}{3+k}-\frac{1}{4+k}\right)+\cdots$$

$$+\left(\frac{1}{s+k}-\frac{1}{s+k+1}\right)$$

$$=\frac{1}{1+k}-\frac{1}{s+k+1}$$

then

$$\lim_{s\to\infty}\sum_{n=1}^{s}\left\{\frac{1}{n+k}-\frac{1}{n+k+1}\right\}=\frac{1}{1+k}$$

and so we arrive at

(6) $$\sum_{n=1}^{\infty}\frac{1}{n(n+1)(n+2)\cdots(n+p)}=\frac{1}{p!}\sum_{k=0}^{p-1}\binom{p-1}{k}(-1)^{k}\frac{1}{1+k}.$$

Now, it is easily checked that[28]

$$\binom{p-1}{k}=\frac{1+k}{p}\binom{p}{1+k}$$

---

28. Just observe that $\binom{p-1}{k}=\frac{(p-1)!}{k!(p-k-1)!}$ and that $\frac{1+k}{p}\binom{p}{1+k}=\frac{1+k}{p}\times\frac{p!}{(1+k)!(p-k-1)!}=\frac{(p-1)!}{k!(p-k-1)!}$.

and so (6) becomes

$$(7) \quad \sum_{n=1}^{\infty} \frac{1}{n(n+1)(n+2)\cdots(n+p)} = \frac{1}{p!} \sum_{k=0}^{p-1} \frac{1+k}{p} \binom{p}{1+k} (-1)^k \frac{1}{1+k}$$

$$= \frac{1}{p} \times \frac{1}{p!} \sum_{k=0}^{p-1} \binom{p}{1+k} (-1)^k.$$

In this last sum, change the index to $j = k + 1$, and so (7) becomes

$$\sum_{n=1}^{\infty} \frac{1}{n(n+1)(n+2)\cdots(n+p)} = \frac{1}{p} \times \frac{1}{p!} \sum_{j=1}^{p} \binom{p}{j} (-1)^{j-1}$$

$$= -\frac{1}{p} \times \frac{1}{p!} \sum_{j=1}^{p} \binom{p}{j} (-1)^{j}$$

$$= -\frac{1}{p} \times \frac{1}{p!} \left[\sum_{j=0}^{p} \binom{p}{j} (-1)^{j} - \binom{p}{0} (-1)^{0} \right] = -\frac{1}{p} \times \frac{1}{p!} \left[\sum_{j=0}^{p} \binom{p}{j} (-1)^{j} - 1 \right].$$

As the final step, notice that

$$\sum_{j=0}^{p} \binom{p}{j} (-1)^{j} = (1-1)^p = 0$$

and so, at last (!),

$$\sum_{n=1}^{\infty} \frac{1}{n(n+1)(n+2)\cdots(n+p)} = \frac{1}{p} \times \frac{1}{p!}.$$

We can numerically check this result (which does have a lot of binomial combinatorial manipulations to it) to ease any concerns.[29] If you did Challenge Problem 1.8.2 by a partial fraction expansion, you know the answer for $p = 3$ is $\frac{1}{18}$, which is indeed the value of $\frac{1}{3} \cdot \frac{1}{3!}$. To explore this numerically, we can easily program the original summation; the *MATLAB* code **inverseprod.m** (in the box) does the job. When run for $p = 3$, using the first 1,000 terms of the sum, the code produced the value of $0.0555555\ldots$, which is in pretty good agreement with $\frac{1}{18} = 0.0555555\ldots$. This agreement is a nice check on the correctness of the code. Our analysis says that the answer for $p = 4$ is $\frac{1}{4} \cdot \frac{1}{4!} = \frac{1}{96} = 0.010416666\ldots$, and the code produced a value of $0.010416666\ldots$, again in excellent agreement with theory.

```
%inverseprod.m
p=input('What is p?')
s=0;
for n=1:1000
    prod=n;
    for loop=1:p
        prod=prod*(n+loop);
    end
    s=s+1/prod;
end
s
```

29. I'm thinking, in particular, of the "physics arguments" I made in claiming that $\binom{a}{b} = 0$ for the two cases $b > a$ and $b < 0$. Mathematicians, in particular, may be less willing to accept such arguments, and in fact, they *do* reject arguing $\binom{a}{b} = 0$ for $a < 0$, a case that for an engineer looks at first blush to also be zero (after all, how many ways can you choose b books from *fewer than none*?). To pursue this would take us too far afield from $\zeta(3)$, but you can find more on how mathematicians handle this situation in my book, *How to Fall Slower Than Gravity* (Princeton University Press, 2018), pp. 34–35, 200–202.

CHAPTER 3

Periodic Functions, Fourier Series, and the Zeta Function

3.1 The Concept of a Function

In this opening section we'll look just a bit deeper at a concept we've taken for granted so far. For a modern mathematician, a function f of a single independent variable t (often thought of as representing time), written as $f(t)$, is a rule that assigns a value to f for each possible value of t. Mathematicians often express this by saying the rule maps t into f. The most common way of defining the mapping rule (that is, the function) is to write an analytical formula: for example, $f(t)=t^2$. But that's not the only possibility. One alternative is to simply write down a column of all possible values of t and then, next to each value in that column, write the associated value of f. This listing might well be infinitely long, but that's okay. We can certainly imagine such a huge list, even if it would take a very long time (and a lot of paper) to actually write it down.

There were analysts in the past who were not terribly enthusiastic about such a liberal view. The French mathematical physicist Jean le Rond d'Alembert (1717–1783), a contemporary of Euler, was of that persuasion, and he championed the strict interpretation that a function absolutely must be expressible via the usual symbols of mathematics. In contrast, Euler was far more liberal and was happy to call $f(t)$ a function if you could simply draw the curve of $f(t)$ versus t. That view does sound plausible (to engineers and physicists, anyway), but there is a troublesome implication tucked away in it that is all too easy to overlook.

Saying that you're "drawing a curve" implies that at almost every instant of drawing time, there is a direction to the motion of the tip of the pen or pencil making the drawing. That is, it is implicitly assumed that the curve has a tangent at nearly every one of its points, which means, in turn, that the curve (function) has a derivative at nearly every point. The "nearly" means there might be a finite number of points of exception. That's because if a curve has a finite number of points at which the derivative fails to exist (for example, $f(t) = |t|$, which obviously fails to have a derivative (tangent) at the single point $t = 0$, as shown in Figure 3.1.1), then we nevertheless can still draw it. Then, in 1861, Riemann cooked up a *continuous* (this is a crucial point, as I'll elaborate on in just a moment) function that he speculated would fail to have a derivative anywhere, even though he could write a simple analytical expression for it.

In (3.1.1) I've written Riemann's function, and Figure 3.1.2 shows three partial sums of (3.1.1) plotted over the interval $0 \le t \le \pi$, and you can see that as more terms are included, the curve does become increasingly "wild." It certainly becomes easy to believe that, as the number of terms goes to infinity, the curve really might fail to have a tangent anywhere. The true state of affairs concerning the differentiability of (3.1.1) was, however, determined only in relatively recent times, and it turns out that Riemann was wrong: His function does have points at which the derivative exists, although it is true

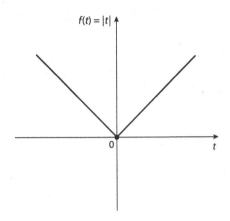

FIGURE 3.1.1.

At $t = 0$, the tangent to $|t|$ has slope ± 1, depending on whether $t = 0$ is approached through positive or negative values of t. This non-uniqueness of the slope at $t = 0$ means $|t|$ has no derivative at that point.

that (3.1.1) is almost everywhere non-differentiable.[1] (Those trying to prove the Riemann hypothesis should keep this partial failure of Riemann's intuition in mind—he was a genius, yes, but he wasn't infallible.)

(3.1.1) $$f(t) = \sum_{n=1}^{\infty} \frac{\sin(n^2 t)}{n^2}.$$

1. Joseph Gerver, "The Differentiability of the Riemann Function at Certain Rational Multiples of π," *American Journal of Mathematics*, January 1970, pp. 33–55. An historical discussion of (3.1.1) is by E. Neuenschwander, "Riemann's Example of a Continuous, 'Nondifferentiable' Function," *Mathematical Intelligencer*, March 1978, pp. 40–44. The concept of a non-differentiable curve has found its way from abstract mathematics into physics: In 1933 the mathematicians (American) Norbert Wiener (1894–1964), (English) Raymond Paley (1907–1933), and (Polish) Antoni Zygmund (1900–1992) showed that the typical continuous path of a particle executing Brownian motion is everywhere non-differentiable.

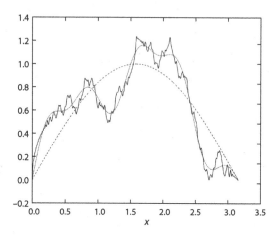

FIGURE 3.1.2.

The partial sums of (3.1.1) for $n = 1$ (dashed), $n = 3$ (dotted), and
$n = 16$ (solid) terms.

Riemann's creation is fun to ponder (with the medical term for
the study of disease in mind, mathematicians somewhat gruesomely
call it a *pathological* function), but it and functions "like it" simply
don't occur in "real life" and so, for engineers and physicists, Euler's
concept of a function as anything you can draw is sufficient for
almost all applications. Mathematicians, of course, are not at all
impressed with such an argument! (In 1872 Weierstrass—do you
recall that he was the fellow who showed us how to go from $n!$ in
(1.4.13) to $(n!)(-n)!$ in (1.4.19)?—finally did conjure up a continuous, nowhere differentiable function.[2]) In 1899 the great French
mathematician Henri Poincaré (1854–1912) correctly wrote of such
bizarre objects like the functions of Riemann and Weierstrass that
"A hundred years ago such a function would have been considered
an outrage on common sense."

2. Weierstrass' function is discussed by E. Hairer and G. Wanner in their elegant
Analysis by Its History (Springer, 1996), pp. 263–269.

Poincaré wrote that just five years before the Swedish mathematician Niels Fabian Helge von Koch (1870–1924) dreamed up yet another continuous two-dimensional function that, like Weierstrass', is nowhere differentiable; it was a creation that would have done nothing to change Poincaré's mind. Unlike the functions of Riemann and Weierstrass, however, where t (time) is the independent variable, von Koch imagined he had the spatial variables x and y of the plane available. And rather than the numerical values of the older functions, von Koch's function value was the *direction* of a curve. What makes von Koch's creation really interesting, however, is that it requires no knowledge of advanced math, like the trigonometric functions that the Riemann and Weierstrass functions use. A bright grammar school student can easily understand von Koch's curve. (In 1915 the Polish mathematician Waclaw Sierpiński (1882–1969) cooked up a similar function, one just slightly more complicated than von Koch's.)

Von Koch's curve starts with a single line segment of unit length and then replaces that line segment with four shorter ones. And then those four line segments are each replaced with four even shorter line segments, and so on, endlessly. If we call the result of the first replacement operation iteration 1, then Figure 3.1.3 shows the first four iterations, with each new iteration obviously more "crinkly" than the last one. It is easy to appreciate, I think, that if one iterates to infinity, the curve becomes so crinkly that there is no direction to the curve at any point on the curve—and yet the curve is continuous everywhere (there are no gaps). Another bizarre property of the von Koch curve is that, as we endlessly iterate, the length of the curve increases without bound (see Challenge Problem 3.1.1), even though the crinkles, individually, become ever smaller. The increasingly lengthy curve nevertheless remains entirely in the same finite region of the plane, as shown in Figure 3.1.3. So, after an infinity of iterations, the von Koch curve can't be drawn for two reasons: (1) it has no direction (tangent) anywhere, and (2) it is infinitely long in a fixed, finite region of space.

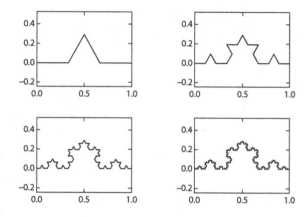

FIGURE 3.1.3.

The first four iterations of the von Koch curve.

It is difficult, after seeing how von Koch's curve is generated, not to think of this posthumously published (1872) jingle by the British mathematician Augustus De Morgan, who, you'll recall from Chapter 2, created the four-color planar map problem:

> Big fleas have little fleas upon their backs to bite 'em,
> And little fleas have lesser fleas, and so, *ad infinitum*.
> And the great fleas, themselves, in turn, have greater fleas to
> go on,
> While these again have greater still, and greater still, *and so on*.

Could that little bit of doggerel have inspired von Koch? Well, who knows—maybe. In fact, however, if von Koch had read the poetry of the Irish writer Jonathan Swift (1667–1745), he would have found that De Morgan had himself been anticipated by more than a century. In a 1733 poem Swift observed how the great and not so great in the world of poetry interact:

So, *naturalists* observe, a flea
Has smaller fleas that on him prey;
And these have smaller still to bite 'em.
And so proceed *ad infinitum*.
Thus every poet, in his kind
Is bit by him that comes behind.

It was De Morgan, however, who almost certainly *did* inspire the American mathematical physicist Lewis Richardson (1881–1953), who in 1922 penned this little rhyme:

Big whorls have little whorls
That feed on their *velocity*;
And little whorls have lesser whorls
And so on to *viscosity*.

The lesson here is clear—if you are looking for inspiration on how to solve a math problem, you might consider seeking it in a study of some aspect of Nature.

As a final comment on the vast possibilities open for something to be called a function, recall that I have emphasized that Riemann's and Weierstrass' functions, like von Koch's, are continuous. That demanding condition is what makes those particular functions so interesting. If, however, one doesn't require continuity, then it is much easier to cook up functions that have no derivative anywhere. Consider, for example, "Dirichlet's function," dating from 1829 (the same Dirichlet who gave his name to the eta function in (2.1.9)):

$$f(t) = \begin{cases} 1, & \text{if } t \text{ is rational} \\ 0, & \text{if } t \text{ is } \textbf{not} \text{ rational} \end{cases}$$

which isn't continuous anywhere. (Try making a sketch of this $f(t)$ as t varies from 0 to 1. I'll bet you can't!) Here's why.

Between any two rational numbers, no matter how close they are, there is another rational number (for example, their average value), and so between 0 and 1, there is an infinite number of rational numbers. And yet, as the Russian-born German mathematician Georg Cantor (1845–1918) showed in 1874, the infinity of the irrationals is greater than that of the rationals![3] That is, we have two infinite sets closely interwoven with each other. So closely, in fact, that it is impossible for continuity to exist. If a function $f(t)$ is continuous at t, then a small deviation from t produces a small deviation in the value of the function. For Dirichlet's function, however, all changes in the function value are 1 in magnitude ($1 \rightarrow 0$ or $0 \rightarrow 1$), and any deviation in t, no matter how small, results in an infinite number of such "not small" changes.

As you probably have gathered by now, the concept of a function is a deep one, and it has been the object of study by some of the greatest mathematicians. For our particular interest in this book, the zeta function, we are going to limit our study to a particular class of functions: *those that are periodic with period T* (terms to be defined in the next section). Such functions have been known since ancient times, but they started to come into their own in the 18th century, from the pens of such masters as Euler (who else!), the Swiss Daniel Bernoulli (1700–1782) who was the son of Euler's mentor in Basel (Johann Bernoulli), and D'Alembert. And then, with the appearance in 1822 of the enormously influential book *The Analytical Theory of Heat* by the French mathematical physicist Joseph Fourier (look back at note 12 in Chapter 2), the use of infinite series whose terms are sinusoidal functions with harmonically related frequencies broke free from pure mathematics and entered the physical

3. For a proof of this not-so-obvious claim (and a discussion of what is meant by saying "one infinity is greater than another infinity"), using only elementary arguments from grammar school arithmetic (!), see my book, *The Logician and the Engineer* (Princeton University Press, 2013), pp. 169–172.

world.[4] With physicists and engineers now in the picture, Fourier series expansions of periodic functions quickly became a tool that no mathematical analyst could be without.

Challenge Problem 3.1.1: How long is the von Koch curve after the nth iteration? (Hint: Using Figure 3.1.3 as a guide, calculate both the number of line segments in the curve after the nth iteration, and the length of each segment.) If the initial line segment (before we start iterating) has a length of 1 inch, how many iterations are required to arrive at a curve with a length that first exceeds 1 light-year? (Take the speed of light as 186,000 miles per second, and use 365 days in a year.) How many line segments are in the curve after the nth iteration?

3.2 Periodic Functions and Their Fourier Series

Fourier's name is today firmly attached to trigonometric series expansions of *periodic* functions. That is, functions that endlessly repeat. If T is the *period* of $f(t)$, which means

$$(3.2.1) \qquad f(t) = f(t+T), \quad -\infty < t < \infty,$$

where T is a finite, positive constant, then we'll assume that we can write

$$(3.2.2) \qquad f(t) = \frac{1}{2}a_0 + \sum_{k=1}^{\infty}\{a_k \cos(k\omega_0 t) + b_k \sin(k\omega_0 t)\},$$

where $\omega_0 = \frac{2\pi}{T}$. The explanation for this relation is that the lowest frequency ($k = 1$) in a Fourier series is associated with the repetition period, T, of each cycle of the periodic function represented by the series. That lowest frequency is $f_0 = \frac{1}{T}$ (measured since 1960 in units

---

4. The story of Fourier series in electrical engineering and thermal physics is told in my book *Hot Molecules, Cold Electrons* (Princeton University Press, 2020). See also my *Transients for Electrical Engineers* (Springer, 2018).

of "hertz"—named after the German mathematical physicist Heinrich Hertz (1857–1894)—which engineers and physicists used to call "cycles per second"), and so the first sinusoidal terms in a Fourier series are $a_1\cos(2\pi f_0 t)$ and/or $b_1\sin(2\pi f_0 t)$. The 2π factor converts the frequency from hertz to *radians per second*. So, $\omega_0 = 2\pi f_0 = \frac{2\pi}{T}$, and subsequent terms in the series after the first term have frequencies that are integer multiples of ω_0, as expressed by (3.2.2).

We've already encountered a Fourier series in this book, of course: As I stated back in (2.3.1), in 1744 Euler declared that

$$(3.2.3) \quad \frac{\pi-t}{2} = \sum_{n=1}^{\infty} \frac{\sin(nt)}{n} = \sin(t) + \frac{\sin(2t)}{2} + \frac{\sin(3t)}{3} + \cdots,$$

which he used to calculate a "cousin" of $\zeta(3)$. This series was probably the very first Fourier series, although Euler didn't call it that, since Fourier wouldn't be born until 24 years later. Euler's series gives the same value as $\frac{\pi-t}{2}$ does, for any t in the interval $0 < t < 2\pi$. (You may be wondering how the left-hand side of (3.2.3) can be called periodic: We'll use Fourier theory to derive (3.2.3) in the next section, and all will be explained there.)

There are two important caveats about T that need to be appreciated. First, T is the smallest possible value for which $f(t)$ repeats (obviously, if $f(t)$ starts to repeat every time t increases by T, it will be repeating for each increase of $2T$, $3T$, and so on). We'll call the smallest repetition time the *fundamental period*. Second, demanding that $T > 0$ eliminates the mathematically trivial case of $f(t)$ a constant being called a periodic function, because in that case, there is no smallest positive T such that $f(t) = f(t + T)$: for every $T > 0$, there is yet a smaller $T > 0$ (for example, the T that is half the previous T). In direct language, a periodic function is a varying function that endlessly repeats itself, in both directions as the independent variable (in our case here, t) goes off to $t = \pm\infty$. This is the mathematician's pure image of a periodic function, which means $f(t)$ must have been "doing its thing" since $t = -\infty$. That means all the periodic functions so beloved by physicists and electrical engineers (like the sinusoidal voltages at the wall outlets of your home) are not truly periodic, because, at some time in the past, they simply didn't yet exist (they had to be turned on). Before that turn-on instant, those voltages were zero, and so $f(t) = f(t + T)$ for all t is just not possible in the physical world. This is a theoretical objection that is, however, routinely ignored by everyone—including mathematicians.

The best-known periodic functions are surely the sinusoidal functions: $f(t) = \sin(t)$ is periodic with period $T = 2\pi$, as is $g(t) = \cos(t)$. It is interesting to note that

$$\frac{f(t)}{g(t)} = \frac{\sin(t)}{\cos(t)} = \tan(t)$$

is also periodic but with a quite different period of π. This result is a hint that, when combining functions that are individually periodic, the result may have a surprise tucked inside. To illustrate what I mean, consider the following two questions as you read the rest of this section, and then I'll ask you about them again as challenge questions.

(a) Suppose $f_1(t)$ and $f_2(t)$ are each periodic, with periods T_1 and T_2, respectively. Is it necessarily true that their sum is periodic? The answer is *no*. Can you think of a counterexample?

(b) Suppose the sum function in (a) is periodic with period T. Is it possible for $T < \min(T_1, T_2)$? That is, can the sum of two periodic functions be periodic with a period less than either of the two original periods? The answer is *yes*. Can you think of a specific example?

Okay, back to Fourier series. If (3.2.2) is to be useful, we obviously have to know what the a_k and b_k coefficients are (the so-called *Fourier coefficients*), and that will be our very first task here, the derivation of expressions for those coefficients. (When we do that, you'll see why that curious $\frac{1}{2}$ in front of a_0 is there.) So, imagine that we want to express $f(t)$, over the symmetrical interval $-\frac{T}{2} < t < \frac{T}{2}$, as a sum of trigonometric terms with frequencies that are multiples of ω_0. That is, for a finite sum of $N + 1$ terms, we write

$$S_N(t) = \frac{1}{2}a_0 + \sum_{k=1}^{N}\left\{a_k\cos(k\omega_0 t) + b_k\sin(k\omega_0 t)\right\}.$$

If we do this, the obvious question now is: What should the a_k and b_k coefficients be to give the best approximation to $f(t)$?

To answer that question, we have to define what is meant by "best." Here's one way to do that. Define the integral

$$J = \int_{-T/2}^{T/2}[f(t) - S_N(t)]^2\,dt,$$

and then ask what the a_k and b_k coefficients should be to minimize J. (J is called the *integrated squared error* of the trigonometric approximation.) If we just calculated the integral of the error alone, without squaring it, we could conceivably get a small J even if there are big positive differences between $f(t)$ and $S_N(t)$, over one or more intervals of t, because they are canceled by big negative differences between $f(t)$ and $S_N(t)$ over other intervals of t. That is, $S_N(t)$ could be greatly different from $f(t)$ for almost all values of t but still result in a small integrated error. By minimizing the integrated *squared* error, however, such cancellations can't occur. For a squared error, a small J forces the series approximation $S_N(t)$ to stay close to $f(t)$ for almost all values of t.

Now,

$$J = \int_{-T/2}^{T/2}\left[f(t) - \left(\frac{1}{2}a_0 + \sum_{k=1}^{N}\left\{a_k\cos(k\omega_0 t) + b_k\sin(k\omega_0 t)\right\}\right)\right]^2\,dt,$$

and we imagine that somehow we have determined the best values for all the a's and all the b's except for one final a (or one final b). For the sake of a specific demonstration, suppose it is a_n that remains to be determined, where n is in the interval 1 to N. From freshman calculus, then, we wish to determine a_n such that

$$\frac{dJ}{da_n} = 0, 1 \leq n \leq N.$$

We'll treat the $n = 0$ case (the value of a_0) separately, later in this section.

Assuming that the derivative of the integral is the integral of the derivative,[5] we have

$$\frac{dJ}{da_n} = \int_{-T/2}^{T/2} 2\left[f(t) - \left(\tfrac{1}{2}a_0 + \Sigma_{k=1}^{N}\{a_k \cos(k\omega_0 t) + b_k \sin(k\omega_0 t)\}\right)\right]\cos(n\omega_0 t)dt$$

and so, setting this equal to zero, we arrive at

$$\int_{-T/2}^{T/2} f(t)\cos(n\omega_0 t)dt$$
$$= \int_{-T/2}^{T/2}\left(\frac{1}{2}a_0 + \sum_{k=1}^{N}\{a_k \cos(k\omega_0 t) + b_k \sin(k\omega_0 t)\}\right)\cos(n\omega_0 t)dt.$$

But since

$$\int_{-T/2}^{T/2} \cos(n\omega_0 t)dt = 0,$$

and since

$$\int_{-T/2}^{T/2} \sin(k\omega_0 t)\cos(n\omega_0 t)dt = 0,$$

5. This is not always true, but it is true for J. There are also details about what mathematicians call *uniform convergence* concerning the J integral, which in the admittedly casual spirit of this book we'll ignore.

and since

$$\int_{-T/2}^{T/2} \cos(k\omega_0 t)\cos(n\omega_0 t)dt = 0 \ \text{ if } \ k \neq n,$$

and since

$$\int_{-T/2}^{T/2} \cos^2(n\omega_0 t)dt = \frac{T}{2},$$

then

$$\int_{-T/2}^{T/2} f(t)\cos(n\omega_0 t)dt = a_n \frac{T}{2}$$

or

(3.2.4) $$a_n = \frac{2}{T}\int_{-T/2}^{T/2} f(t)\cos(n\omega_0 t)dt, \ 1 \leq n \leq N.$$

By the same argument,

(3.2.5) $$b_n = \frac{2}{T}\int_{-T/2}^{T/2} f(t)\sin(n\omega_0 t)dt, \ 1 \leq n \leq N.$$

Since we've made no special assumptions about n, then (3.2.4) and (3.2.5) hold for any n in the interval $1 \leq n \leq N$. Notice that the Fourier coefficients have no dependence on N.

Now, what about a_0 (the $n = 0$ coefficient)? We have

$$\frac{dJ}{da_0} = \int_{-T/2}^{T/2} 2\left[f(t) - \left(\frac{1}{2}a_0 + \sum_{k=1}^{N}\{a_k\cos(k\omega_0 t) + b_k\sin(k\omega_0 t)\} \right)\right]\left(-\frac{1}{2}\right)dt = 0$$

and so

$$\int_{-T/2}^{T/2} f(t)dt = \int_{-T/2}^{T/2} \frac{1}{2}a_0 dt = \frac{1}{2}Ta_0$$

or

$$(3.2.6) \qquad a_0 = \frac{2}{T} \int_{-T/2}^{T/2} f(t)dt.$$

By including the $\frac{1}{2}$ factor in front of a_0 in (3.2.2), we have arrived at an expression for a_0 that is correctly given by (3.2.4), the expression for a_n, $1 \le n \le N$, even when we set $n = 0$ in (3.2.4), and that is why mathematicians write $\frac{1}{2}a_0$ instead of just a_0. It's a matter of elegance. (The physical significance of $\frac{1}{2}a_0$ is the average value of $f(t)$ over a period.)

There is a powerful mathematical theorem that says, for the coefficients we have just calculated, not only is J minimized, but also the integrated squared error actually goes to zero as $N \to \infty$ in S_N, *as long as $f(t)$ has just a finite number of discontinuities in a period*. That is a physical requirement that is certainly satisfied in any real-world problem.

Observe that a series will, in general, contain both sine and cosine terms. There are, however, certain special (but highly useful) functions whose Fourier series have only sines (or only cosines). First, suppose $f(t)$ is an *even* function over the interval $-\frac{T}{2} < t < \frac{T}{2}$. That is, $f(-t) = f(t)$. Then the integrand in (3.2.4) is even (because $\cos(n\omega_0 t)$ is even) and, for functions, even times even is even, while the integrand in (3.2.5) is odd (because $\sin(n\omega_0 t)$ is odd) and, for functions, even times odd is odd. Thus, while a_n will in general be non-zero, all the b_n coefficients will vanish, and so the Fourier series for an even $f(t)$ will have only cosine terms. In contrast, suppose $f(t)$ is an odd function over the interval $-\frac{T}{2} < t < \frac{T}{2}$. That is, $f(-t) = -f(t)$. Now the opposite situation results and, while in general the b_n will be non-zero, all the a_n coefficients will vanish, and so the Fourier series for an odd $f(t)$ will have only sine terms.[6]

6. It's a curious fact that, while the conditions of evenness and oddness are quite restrictive, *any* function can always be written as the sum of an even function and an odd function. The proof is easy, direct, and convincing, as it's a proof by construction (the best kind of all). First, whatever $f(t)$ is, $f(t) + f(-t)$ is even, and $f(t) - f(-t)$ is odd. Then simply observe that $f(t) = \frac{1}{2}[f(t) + f(-t)] + \frac{1}{2}[f(t) - f(-t)]$. Done!

When we evaluate the Fourier series of an $f(t)$ defined on the interval $-\frac{T}{2} < t < \frac{T}{2}$, the result will *not* equal $f(\hat{t})$ for a $\hat{t}$ outside the interval $-\frac{T}{2}$ to $\frac{T}{2}$. The series will indeed converge for that $\hat{t}$, but not to $f(\hat{t})$ but rather to the value of what is called the *periodic extension* of $f(t)$ up and down the t-axis. Figure 3.2.1 shows the periodic extension (for $T = 2\pi$) of an even function (t^2), and Figure 3.2.2 shows the periodic extension (for $T = 2\pi$) of an odd function (t).

Challenge Problem 3.2.1: Take a look back at the two questions about periodic functions that I asked you (just after the last box) to think about. What have you concluded?

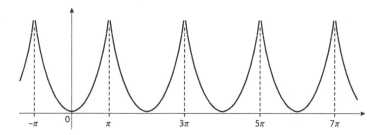

FIGURE 3.2.1.

The periodic extension of an even function.

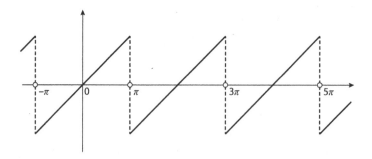

FIGURE 3.2.2.

The periodic extension of an odd function.

3.3 Complex Fourier Series and Parseval's Power Formula

To demonstrate the power of Fourier series, in this section I'll take you through two detailed problems. Both will demonstrate the use of Euler's identity in developing a complex form of Fourier series. To set you up for these calculations, it is convenient to do some preliminary (but still pretty straightforward) analysis. Repeating (3.2.2),

$$f(t) = \frac{1}{2}a_0 + \sum_{k=1}^{\infty}\{a_k \cos(k\omega_0 t) + b_k \sin(k\omega_0 t)\}$$

where $\omega_0 T = 2\pi$. Using Euler's identity (remember, $i = \sqrt{-1}$), this becomes

$$f(t) = \frac{1}{2}a_0 + \sum_{k=1}^{\infty}\left\{a_k \frac{e^{ik\omega_0 t} + e^{-ik\omega_0 t}}{2} + b_k \frac{e^{ik\omega_0 t} - e^{-ik\omega_0 t}}{2i}\right\}$$

$$= \frac{1}{2}a_0 + \sum_{k=1}^{\infty}\left[\left\{\frac{a_k}{2} + \frac{b_k}{2i}\right\}e^{ik\omega_0 t} + \left\{\frac{a_k}{2} - \frac{b_k}{2i}\right\}e^{-ik\omega_0 t}\right].$$

If we let the summation index run from minus to plus infinity, then we can write a *complex* Fourier series as

$$(3.3.1) \qquad f(t) = \sum_{k=-\infty}^{\infty} c_k e^{ik\omega_0 t}, \quad \omega_0 T = 2\pi$$

where the c_k are constants (in general, complex-valued constants).

Let's next suppose that $f(t)$ is a real-valued function. Since the conjugate of a real value is the real value, then

$$f(t) = \sum_{k=-\infty}^{\infty} c_k e^{ik\omega_0 t} = f^*(t) = \left\{\sum_{k=-\infty}^{\infty} c_k e^{ik\omega_0 t}\right\}^*.$$

Since the conjugate of a sum is the sum of the conjugates, and since the conjugate of a product is the product of the conjugates (you should confirm these claims), then

(3.3.2) $$\sum_{k=-\infty}^{\infty} c_k e^{ik\omega_0 t} = \sum_{k=-\infty}^{\infty} c_k^* e^{-ik\omega_0 t},$$

which tells us that, if $f(t)$ is a real-valued function, then $c_{-k} = c_k^*$ (to see this, set the coefficients of matching exponential terms on each side of (3.3.2) equal to each other). Notice, too, that for the case of $k = 0$ we have $c_0 = c_0^*$, which says that, for any real-valued function $f(t)$, we'll have c_0 always come out as real valued.

Now, for some (any) particular integer from minus to plus infinity (let's say, n), multiply both sides of (3.3.1) by $e^{-in\omega_0 t}$ and integrate over a period, that is, over any interval of length T. Then, with t' an arbitrary (but fixed) value of t,

$$\int_{t'}^{t'+T} f(t)e^{-in\omega_0 t} dt = \int_{t'}^{t'+T} \left\{ \sum_{k=-\infty}^{\infty} c_k e^{ik\omega_0 t} \right\} e^{-in\omega_0 t} dt$$

or

(3.3.3) $$\int_{t'}^{t'+T} f(t)e^{-in\omega_0 t} dt = \sum_{k=-\infty}^{\infty} c_k \int_{t'}^{t'+T} e^{i(k-n)\omega_0 t} dt.$$

The integral on the right in (3.3.3) is easy to do, and we'll do it in two steps. Once for $k \neq n$, and then again for $k = n$. So, if $k \neq n$ we have

$$\int_{t'}^{t'+T} e^{-i(k-n)\omega_0 t} dt = \left\{ \frac{e^{i(k-n)\omega_0 t}}{i(k-n)\omega_0} \right\} \Big|_{t'}^{t'+T} = \frac{e^{i(k-n)\omega_0 (t'+T)} - e^{i(k-n)\omega_0 t'}}{i(k-n)\omega_0}$$

$$= \frac{e^{i(k-n)\omega_0 t'} \{ e^{i(k-n)\omega_0 T} - 1 \}}{i(k-n)\omega_0}.$$

Since $\omega_0 T = 2\pi$, since $k - n$ is a non-zero integer, and since Euler's identity tells us that $e^{i(k-n)\omega_0 T} = 1$, then the integral is zero for the $k \neq n$ case. For the $k = n$ case, the integral becomes

$$\int_{t'}^{t'+T} e^0 dt = \{t\} \big|_{t'}^{t'+T} = T.$$

So, in summary,

(3.3.4)
$$\int_{period} e^{i(k-n)\omega_0 t} dt = \begin{cases} 0, \, k \neq n \\ T, \, k = n \end{cases}.$$

Thus, (3.3.3) becomes

$$\int_{period} f(t) e^{-in\omega_0 t} dt = c_n T$$

or, for all k, the Fourier coefficients in (3.3.1) are given by (just replace n with k on both sides of the last integral)

(3.3.5)
$$c_k = \frac{1}{T} \int_{period} f(t) e^{-ik\omega_0 t} dt, \; \omega_0 T = 2\pi.$$

Okay, that's the end of the preliminary analysis I mentioned. Now, let's use it to derive Euler's Fourier series of (3.2.3).

We start by writing, as in (3.3.1), Euler's function as

(3.3.6)
$$f(t) = \frac{\pi - t}{2} = \sum_{k=-\infty}^{\infty} c_k e^{ik\omega_0 t}, \; 0 < t < 2\pi.$$

We'll take one period to be $T = 2\pi$, and so the condition $\omega_0 T = 2\pi$ means that $\omega_0 = 1$. From (3.3.5), we have

$$c_k = \frac{1}{2\pi} \int_0^{2\pi} \frac{\pi - t}{2} e^{-ikt} dt$$

or,

$$(3.3.7) \qquad c_k = \frac{1}{4}\int_0^{2\pi} e^{-ikt}\,dt - \frac{1}{4\pi}\int_0^{2\pi} te^{-ikt}\,dt.$$

The case of $k = 0$ is easy to do:

$$c_0 = \frac{1}{4}\int_0^{2\pi} dt - \frac{1}{4\pi}\int_0^{2\pi} t\,dt = \frac{1}{4}(t)\big|_0^{2\pi} - \frac{1}{4\pi}\left(\frac{t^2}{2}\right)\big|_0^{2\pi} = \frac{2\pi}{4} - \frac{4\pi^2}{8\pi}$$

and so

$$(3.3.8) \qquad\qquad\qquad c_0 = 0.$$

For the case of $k \neq 0$, the first integral in (3.3.7) is

$$\frac{1}{4}\int_0^{2\pi} e^{-ikt}\,dt = \frac{1}{4}\left(\frac{e^{-ikt}}{-ik}\right)\big|_0^{2\pi} = \frac{e^{-i2\pi k} - 1}{-i4k} = 0$$

because $e^{-i2\pi k} = 1$ for any integer k. Thus, our result for $c_{k\neq0}$ reduces to

$$c_{k\neq0} = -\frac{1}{4\pi}\int_0^{2\pi} te^{-ikt}\,dt.$$

Using integration by parts (or a good table of integrals) we find, for any constant $a \neq 0$, that

$$\int te^{at}\,dt = \frac{e^{at}}{a}\left(t - \frac{1}{a}\right).$$

Setting $a = -ik$, we have

$$c_{k \neq 0} = -\frac{1}{4\pi} \left\{ \frac{e^{-ikt}}{-ik} \left(t - \frac{1}{-ik} \right) \right\} \Big|_0^{2\pi} = -\frac{i}{4\pi k} \left\{ e^{-ikt} \left(t - \frac{i}{k} \right) \right\} \Big|_0^{2\pi}$$

$$= -\frac{i}{4\pi k} \left\{ e^{-ik2\pi} \left(2\pi - \frac{i}{k} \right) + \frac{i}{k} \right\}$$

or, as $e^{-ik2\pi} = 1$, we have

$$c_{k \neq 0} = -\frac{i}{4\pi k} \left\{ \left(2\pi - \frac{i}{k} \right) + \frac{i}{k} \right\} = -\frac{i}{2k}.$$

Putting this last result and (3.3.8) into (3.3.6), with $\omega_0 = 1$, we have

$$\frac{\pi - t}{2} = \sum_{k=-\infty, k \neq 0}^{\infty} -\frac{i}{2k} e^{ikt}, \ 0 < t < 2\pi$$

or, writing the summation out in pairs of terms ($k = \pm 1, \pm 2, \pm 3, \ldots$), we arrive at

$$\frac{\pi - t}{2} = -\frac{i}{2} \left[\left(\frac{e^{it} - e^{-it}}{1} \right) + \left(\frac{e^{i2t} - e^{-i2t}}{2} \right) + \left(\frac{e^{i3t} - e^{-i3t}}{3} \right) + \cdots \right]$$

$$= -\frac{i}{2} \left[2i\sin(t) + \frac{2i\sin(2t)}{2} + \frac{2i\sin(3t)}{3} + \cdots \right] = \sin(t) + \frac{\sin(2t)}{2} + \frac{\sin(3t)}{3} + \cdots$$

which is Euler's series from 1744. Notice, in particular, that if we set $t = \frac{\pi}{2}$, Euler's series reduces to

$$\frac{\pi}{4} = 1 - \frac{1}{3} + \frac{1}{5} - \frac{1}{7} + \cdots,$$

a famous result found (via other means) by the Scottish mathematician James Gregory (1638–1675) and the German mathematician

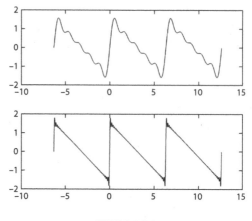

FIGURE 3.3.1.

Euler's "first Fourier series" with five terms (top) and with 40 terms (bottom).

Leibniz. The series on the right is worthless (in a practical sense) for calculating π, because it converges so slowly, but there can be no denying it is a beautiful expression.[7]

Let me now do something you may have already been wondering about, ever since I first wrote (2.3.1): What does a plot of Euler's series actually look like? The claim is that it equals $\frac{\pi-t}{2}$ in the interval $0 < t < 2\pi$, but does it? Figure 3.3.1 shows two plots of Euler's series, each with a different number of terms, as t varies over three full periods. You can see the series does indeed appear to become an increasingly better approximation to $\frac{\pi-t}{2}$ (a downward-sloping ramp), for $0 < t < 2\pi$, as the number of terms increases.[8]

---

7. It was discoveries like this (and the series for ln(2) in (1.3.5)) that inspired the German mathematician Leopold Kronecker (1823–1891) to declare "God made the integers, all else is the work of man."

8. The wiggles in the middle of each period are due to the finite number of terms used, and they decrease in amplitude as the number of terms increases. The curious wiggles in the neighborhoods of the start and end of each period, however, have a different story, one I won't pursue here except to say those wiggles never disappear, even as the number of terms increases without bound. If you want to read more about that phenomenon, take a look at *Dr. Euler* (see note 13, Chapter 2), on that book's pp. 171–173.

To complete this section, here's the second detailed analysis I promised you that demonstrates the use of the complex form of Fourier series, a demonstration that will provide us with yet another derivation of $\zeta(2) = \frac{\pi^2}{6}$. We start by defining the *energy* (a term used by mathematicians, physicists, and engineers alike) of the real-valued periodic function $f(t)$, over a period, to be the integral

$$(3.3.9) \qquad W = \int_{period} f^2(t)dt.$$

(The reason for calling W the energy of $f(t)$ is that if $f(t)$ is a periodic voltage drop across a 1-ohm resistor, then W is the electrical energy dissipated—as heat—by the resistor during one period. Alternatively, if $f(t)$ is a periodic current in a 1-ohm resistor then, again, W is the electrical energy dissipated as heat by the resistor during one period. If you're not into electrical engineering or physics, ignore this business of $f(t)$ being a voltage drop or a current and just take (3.3.9) as a definition.) If we substitute the complex Fourier series for $f(t)$ into the energy integral, writing $f^2(t) = f(t)f(t)$ and using a different index of summation for each $f(t)$, we get

$$W = \int_{period} \left\{ \sum_{m=-\infty}^{\infty} c_m e^{im\omega_0 t} \right\} \left\{ \sum_{n=-\infty}^{\infty} c_n e^{in\omega_0 t} \right\} dt$$

or

$$(3.3.10) \qquad W = \sum_{m=-\infty}^{\infty} \sum_{n=-\infty}^{\infty} c_m c_n \int_{period} e^{i(m+n)\omega_0 t} dt.$$

This last integral is one we've already done, back in (3.3.4). That is, the integral is zero when $m + n \neq 0$ and is T when $m + n = 0$ (when $m = -n$). So, remembering that for a real-valued $f(t)$, $c_{-k} = c_k^*$, we see that (3.3.10) reduces to

$$W = \sum_{k=-\infty}^{\infty} c_k c_{-k} T = T \sum_{k=-\infty}^{\infty} c_k c_k^* = T \sum_{k=-\infty}^{\infty} |c_k|^2$$

or, remembering that c_0 is real since $f(t)$ is real,

$$(3.3.11) \qquad \frac{W}{T} = \frac{1}{T} \int_{period} f^2(t)dt = c_0^2 + 2\sum_{k=1}^{\infty} |c_k|^2,$$

a result called *Parseval's power formula*.[9] (Energy per unit time, the units of $\frac{W}{T}$, is *power* in physics lingo.)

As a spectacular example of the utility of complex Fourier series, here's a derivation of a result we'll use in the next section to calculate $\zeta(2n)$ for n any positive integer. We start with $f(t) = \cos(at)$, $-\pi < t < \pi$, where a is any real, *non*-integer (I'll tell you, soon, why we impose this curious restriction). If we next imagine periodically extending $f(t)$ over the entire t-axis, with period $T = 2\pi$ ($\omega_0 = 1$), then we can write

$$(3.3.12) \qquad f(t) = \sum_{k=-\infty}^{\infty} c_k e^{ikt}$$

where, as we carefully work our way through the detailed algebra, we have

$$c_k = \frac{1}{2\pi}\int_{-\pi}^{\pi} \cos(\alpha t)e^{-ikt}dt = \frac{1}{2\pi}\int_{-\pi}^{\pi} \frac{e^{i\alpha t}+e^{-i\alpha t}}{2}e^{-ikt}dt$$

$$= \frac{1}{4\pi}\left[\int_{-\pi}^{\pi} e^{i(\alpha-k)t}dt + \int_{-\pi}^{\pi} e^{-i(\alpha+k)t}dt\right]$$

9. After the French mathematician Marc-Antoine Parseval des Chenes (1755–1836), who published his formula (in much different form, in a non-Fourier series context) in 1799. There is a most interesting conclusion that immediately follows from Parseval's formula if we make the physically plausible assumption that $f(t)$ has finite energy in a period. Since it is a necessary condition for the convergence of the sum on the right in (3.3.11) that the terms of the sum go to zero as $k \to \infty$ (that is, $\lim_{k\to\infty} c_k = 0$), we see that as frequency increases, the amplitudes of the Fourier coefficients must approach zero. This almost sounds like the result of a physics experiment, but it is purely mathematical in origin.

$$= \frac{1}{4\pi}\left[\frac{e^{i(\alpha-k)t}}{i(\alpha-k)} + \frac{e^{-i(\alpha+k)t}}{-i(\alpha+k)}\right]\Big|_{-\pi}^{\pi}$$

$$= \frac{1}{4\pi i}\left[\frac{e^{i(\alpha-k)\pi} - e^{-i(\alpha-k)\pi}}{\alpha-k} - \frac{e^{-i(\alpha+k)\pi} - e^{i(\alpha+k)\pi}}{\alpha+k}\right]$$

$$= \frac{1}{4\pi i}\left[\frac{\alpha e^{i(\alpha-k)\pi} - \alpha e^{-i(\alpha-k)\pi} + k e^{i(\alpha-k)\pi} - k e^{-i(\alpha-k)\pi}}{\alpha^2 - k^2}\right]$$

$$+ \frac{1}{4\pi i}\left[\frac{-\alpha e^{-i(\alpha+k)\pi} + \alpha e^{i(\alpha+k)\pi} + k e^{-i(\alpha+k)\pi} - k e^{i(\alpha+k)\pi}}{\alpha^2 - k^2}\right]$$

$$= \frac{1}{4\pi i}\left[\frac{\{\alpha e^{i(\alpha-k)\pi} + \alpha e^{i(\alpha+k)\pi}\} + \{-\alpha e^{-i(\alpha-k)\pi} - \alpha e^{-i(\alpha+k)\pi}\}}{\alpha^2 - k^2}\right]$$

$$+ \frac{1}{4\pi i}\left[\frac{\{k e^{i(\alpha-k)\pi} - k e^{i(\alpha+k)\pi}\} + \{k e^{-i(\alpha+k)\pi} - k e^{-i(\alpha-k)\pi}\}}{\alpha^2 - k^2}\right]$$

$$= \frac{1}{4\pi i}\left[\frac{\alpha e^{i\alpha\pi}(e^{-ik\pi} + e^{ik\pi}) - \alpha e^{-i\alpha\pi}(e^{ik\pi} + e^{-ik\pi})}{\alpha^2 - k^2}\right]$$

$$+ \frac{1}{4\pi i}\left[\frac{k e^{i\alpha\pi}(e^{-ik\pi} - e^{ik\pi}) + k e^{-i\alpha\pi}(e^{-ik\pi} - e^{ik\pi})}{\alpha^2 - k^2}\right]$$

$$= \frac{1}{4\pi i}\left[\frac{\alpha e^{i\alpha\pi} 2\cos(k\pi) - \alpha e^{-i\alpha\pi} 2\cos(k\pi)}{\alpha^2 - k^2}\right]$$

$$+ \frac{1}{4\pi i}\left[\frac{-k e^{i\alpha\pi} 2i\sin(k\pi) - k e^{-i\alpha\pi} 2i\sin(k\pi)}{\alpha^2 - k^2}\right]$$

or, as $\sin(k\pi) = 0$ for all integer k,

$$c_k = \frac{2\alpha \cos(k\pi)(e^{i\alpha\pi} - e^{-i\alpha\pi})}{4\pi i(\alpha^2 - k^2)} = \frac{2\alpha \cos(k\pi)2i\sin(\alpha\pi)}{4\pi i(\alpha^2 - k^2)}$$

or, as $\cos(k\pi) = (-1)^k$,

$$(3.3.13) \qquad\qquad c_k = \frac{\alpha(-1)^k \sin(\alpha\pi)}{\pi(\alpha^2 - k^2)}.$$

Substituting (3.3.13) into (3.3.12), we have

$$f(t) = \cos(\alpha t) = \sum_{k=-\infty}^{\infty} \frac{\alpha(-1)^k \sin(\alpha\pi)}{\pi(\alpha^2 - k^2)} e^{ikt}$$

or, if we write the $k = 0$ term separately,

$$\cos(\alpha t) = \frac{\sin(\alpha\pi)}{\pi\alpha} + \sum_{k=1}^{\infty} \frac{\alpha(-1)^k \sin(\alpha\pi)}{\pi(\alpha^2 - k^2)} \{e^{ikt} + e^{-ikt}\}$$

$$= \frac{\sin(\alpha\pi)}{\pi\alpha} + \frac{\alpha\sin(\alpha\pi)}{\pi} \sum_{k=1}^{\infty} \frac{(-1)^k}{\pi(\alpha^2 - k^2)} 2\cos(kt)$$

or, at last,

$$(3.3.14) \qquad \cos(\alpha t) = \frac{\sin(\alpha\pi)}{\pi}\left[\frac{1}{\alpha} + 2\alpha \sum_{k=1}^{\infty} \frac{(-1)^k}{\alpha^2 - k^2} \cos(kt)\right].$$

You should by now have seen why we excluded the case of α being an integer: If it were an integer, then when the summation index reached that integer, the denominator of (3.3.14) would be zero and the sum would blow up. For any non-integer a we avoid that catastrophic event. We'll return to (3.3.14) in the next section.

As a final, dramatic illustration of the use of (3.3.11), suppose that $f(t) = e^{-pt}$, $0 < t < T = 2\pi$, p a non-negative but otherwise arbitrary constant, which defines a single period of a function extended over the entire t-axis. Then, from (3.3.5), and observing that $\omega_0 = 1$,

$$c_k = \frac{1}{2\pi}\int_0^{2\pi} e^{-pt} e^{-ikt} dt = \frac{1}{2\pi}\int_0^{2\pi} e^{-(p+ik)t} dt = \frac{1}{2\pi}\left\{\frac{e^{-(p+ik)t}}{-(p+ik)}\right\}\Big|_0^{2\pi}$$

$$= \left(\frac{1}{2\pi}\right)\frac{1-e^{-(p+ik)2\pi}}{p+ik} = \left(\frac{1}{2\pi}\right)\frac{1-e^{-2\pi p}e^{-ik2\pi}}{p+ik} = \frac{1-e^{-2\pi p}}{2\pi(p+ik)}$$

because $e^{-ik2\pi} = 1$ for all integer k. Thus,

$$|c_k|^2 = \frac{(1-e^{-2\pi p})^2}{4\pi^2(p^2+k^2)}, \quad c_0 = \frac{1-e^{-2\pi p}}{2\pi p}$$

where the expression for c_0 follows because $1 - e^{-2\pi p} > 0$ since $p > 0$. Therefore, (3.3.11) says

$$\frac{1}{2\pi}\int_0^{2\pi} e^{-2pt} dt = \left(\frac{1-e^{-2\pi p}}{2\pi p}\right)^2 + 2\sum_{k=1}^{\infty}\frac{(1-e^{-2\pi p})^2}{4\pi^2(p^2+k^2)}$$

or

(3.3.15) $$\frac{1}{2\pi(1-e^{-2\pi p})^2}\int_0^{2\pi} e^{-2pt} dt = \frac{1}{4\pi^2 p^2} + \frac{1}{2\pi^2}\sum_{k=1}^{\infty}\frac{1}{p^2+k^2}.$$

The integral on the left in (3.3.15) is easy to do:

$$\int_0^{2\pi} e^{-2pt}\,dt = \left(\frac{e^{-2pt}}{-2p}\right)\Big|_0^{2\pi} = \frac{1-e^{-4\pi p}}{2p} = \frac{(1-e^{-2\pi p})(1+e^{-2\pi p})}{2p}$$

and so (3.3.15) becomes

$$\frac{1+e^{-2\pi p}}{4\pi p(1-e^{-2\pi p})} = \frac{1}{4\pi^2 p^2} + \frac{1}{2\pi^2}\sum_{k=1}^{\infty}\frac{1}{p^2+k^2}$$

or

$$\sum_{k=1}^{\infty}\frac{1}{p^2+k^2} = 2\pi^2\left[\frac{1+e^{-2\pi p}}{4\pi p(1-e^{-2\pi p})} - \frac{1}{4\pi^2 p^2}\right]$$

or, at last,

(3.3.16) $$\sum_{k=1}^{\infty}\frac{1}{p^2+k^2} = \left(\frac{\pi}{2p}\right)\frac{1+e^{-2\pi p}}{1-e^{-2\pi p}} - \frac{1}{2p^2}.$$

You'll notice that as p decreases toward zero, the left-hand side of (3.3.16) approaches the value of $\zeta(2)$. This suggests that

(3.3.17) $$\lim_{p\to 0}\left\{\left(\frac{\pi}{2p}\right)\frac{1+e^{-2\pi p}}{1-e^{-2\pi p}} - \frac{1}{2p^2}\right\} = \zeta(2) = \frac{\pi^2}{6}.$$

Challenge Problem 3.3.1: Sum the series $\sum_{k=1}^{\infty}\frac{1}{1-4k^2} = -\frac{1}{3}-\frac{1}{15}-\frac{1}{35}$ $-\frac{1}{63}-\cdots = ?$ in two different ways. Hint: For one way, set $t = \pi$ and $\alpha = \frac{1}{2}$ in (3.3.14). Can you think of a more direct analysis, one that doesn't use the enormous power of Fourier theory?

Challenge Problem 3.3.2: Prove (3.3.17), that $\lim_{p\to 0}\{(\frac{\pi}{2p})\frac{1+e^{-2\pi p}}{1-e^{-2\pi p}}$ $-\frac{1}{2p^2}\} = \frac{\pi^2}{6}$. Hint: Start by making a power series expansion of $e^{-2\pi p}$ and retain the terms up to the one in p^3.

Challenge Problem 3.3.3: In note 20 of Chapter 2 I argued that the Riemann-Lebesgue lemma, which says that for any well-behaved, physically reasonable $f(t)$ defined over the interval $a < t < b$, we have $\lim_{m\to\infty} \int_a^b f(t)\cos(mt)dt = \lim_{m\to\infty} \int_a^b f(t)\sin(mt)dt = 0$, and that an area interpretation of the integrals makes the lemma plausible. Then, in note 9 of this chapter, it is observed that the convergence of the sum in (3.3.11) says that the Fourier coefficients of a periodically extended $f(t)$ must go to zero as frequency increases. Explain how the second note provides alternative support for the lemma.

3.4 Calculating ζ (2n) with Fourier Series

We ended the last section with the calculation of $\zeta(2)$ using Fourier series. Our calculation was, in fact, a special case of a more general analysis but, if we are willing to give up the generality, then $\zeta(2)$ can be directly (and quickly) calculated in just a few easy steps. Furthermore, this approach extends in an obvious way to the calculation of $\zeta(2n)$ for any n, not just for $n = 1$.

We start by defining

$$(3.4.1) \qquad f(t) = t^2, \; -\pi \le t \le \pi,$$

and then periodically extending (3.4.1) over the entire t-axis. That is, we are now working with an even periodic function with period $T = 2\pi$ (and so $\omega_0 = 1$). (Take a look back at Figure 3.2.1.) Because the extended function is even, we know its Fourier series will contain only cosine terms, and so the Fourier coefficients b_N will vanish for all n. That is, we need only to calculate the coefficients a_0 and $a_{n \ne 0}$, and then substitute the results into (3.2.2). So, from (3.2.6) we have

$$(3.4.2) \qquad a_0 = \frac{1}{\pi}\int_{-\pi}^{\pi} t^2 dt = \frac{1}{\pi}\left(\frac{1}{3}t^3\right)\bigg|_{-\pi}^{\pi} = \frac{2}{3}\pi^2.$$

Also, from (3.2.4) we have

$$a_{n\neq0} = \frac{1}{\pi}\int_{-\pi}^{\pi} t^2 \cos(nt)dt = \frac{1}{\pi}\left\{ t^2\frac{\sin(nt)}{n} + 2t\frac{\cos(nt)}{n^2} - 2\frac{\sin(nt)}{n^3}\right\}\bigg|_{-\pi}^{\pi}$$

or

(3.4.3) $$a_{n\neq0} = \frac{4}{n^2}\cos(n\pi).$$

Putting (3.4.2) and (3.4.3) into (3.2.2) gives us

(3.4.4) $$t^2 = \frac{1}{3}\pi^2 + 4\sum_{n=1}^{\infty}\frac{\cos(n\pi)\cos(nt)}{n^2}, -\pi \leq t \leq \pi.$$

Be careful to particularly note that the limits on t include the endpoints of the interval $-\pi$ to π, because the periodically extended function is everywhere continuous (take another look back at Figure 3.2.1). This will prove, in our later discussions, to be of crucial significance.

Now, for the quick conclusion to our calculations: Simply set $t = \pi$ in (3.4.4). That gives

$$\pi^2 = \frac{1}{3}\pi^2 + 4\sum_{n=1}^{\infty}\frac{\cos^2(n\pi)}{n^2} = \frac{1}{3}\pi^2 + 4\sum_{n=1}^{\infty}\frac{1}{n^2}$$

or

(3.4.5) $$\frac{2}{3}\pi^2 = 4\zeta(2),$$

which instantly gives us $\zeta(2) = \frac{\pi^2}{6}$.

That was certainly easy and, in fact, this approach works just as smoothly for $\zeta(4)$, $\zeta(6)$, $\zeta(8)$, and so on. For example, here's the calculation for $\zeta(4)$, and be sure to notice how every step mirrors what we did for $\zeta(2)$. We start by defining

(3.4.6) $$f(t)=t^4, -\pi \le t \le \pi$$

and then periodically extend this over the entire t-axis. As before, the resulting periodic function is even, with period $T = 2\pi$ ($\omega_0 = 1$), and everywhere continuous. And, as before, we need calculate only a_0 and $a_{n \ne 0}$. So,

(3.4.7) $$a_0 = \frac{1}{\pi} \int_{-\pi}^{\pi} t^4 dt = \frac{1}{\pi} \left(\frac{1}{5} t^5 \right) \Big|_{-\pi}^{\pi} = \frac{2}{5} \pi^4.$$

Also,

$$a_{n \ne 0} = \frac{1}{\pi} \int_{-\pi}^{\pi} t^4 \cos(nt) dt$$

$$= \frac{1}{\pi} \left\{ \frac{4t^3 \cos(nt)}{n^2} - \frac{24t \cos(nt)}{n^4} + \frac{t^4 \sin(nt)}{n} - \frac{12t^2 \sin(nt)}{n^3} + \frac{24 \sin(nt)}{n^5} \right\} \Big|_{-\pi}^{\pi}$$

or

(3.4.8) $$a_{n \ne 0} = \left\{ \frac{8\pi^2}{n^2} - \frac{48}{n^4} \right\} \cos(n\pi).$$

Thus,

(3.4.9) $$t^4 = \frac{1}{5} \pi^4 + \sum_{n=1}^{\infty} \left\{ \frac{8\pi^2}{n^2} - \frac{48}{n^4} \right\} \cos(n\pi) \cos(nt), -\pi \le t \le \pi.$$

Setting $t = \pi$ in (3.4.9), we get

$$\pi^4 = \frac{1}{5}\pi^4 + 8\pi^2 \sum_{n=1}^{\infty} \frac{\cos^2(n\pi)}{n^2} - 48\sum_{n=1}^{\infty} \frac{\cos^2(n\pi)}{n^4}$$

or

$$\frac{4}{5}\pi^4 = 8\pi^2\zeta(2) - 48\zeta(4),$$

which becomes

$$\zeta(4) = \frac{8\pi^2\zeta(2) - \dfrac{4}{5}\pi^4}{48} = \frac{8\pi^2\dfrac{\pi^2}{6} - \dfrac{4}{5}\pi^4}{48}.$$

Some easy arithmetic quickly gives us our answer:

(3.4.10) $$\zeta(4) = \frac{\pi^4}{90}.$$

Rather than doing a separate analysis for each $\zeta(2n)$, there is an elegant procedure for calculating all the $\zeta(2n)$ at once. The first step is easy: just look up the power series expansion of $\cot(x)$ about $x = 0$ (the *Taylor series*, after the English mathematician Brook Taylor (1685–1731), who published in 1715 but had been anticipated *by decades* by James Gregory, who you'll recall discovered a famous series for $\frac{\pi}{4}$) in any good set of math tables, where you'll find[10]

$$\cot(x) = \frac{1}{x} - \frac{1}{3}x - \frac{1}{45}x^3 - \frac{2}{945}x^5 - \frac{1}{4,725}x^7 - \cdots,$$

---

10. Calculating the Taylor series of a function $f(x)$ is a routine exercise in freshman calculus, and I'll let you look up the details in any good textbook if you need to refresh your memory. Purists will say that the Taylor series around $x = 0$ is better called a Maclaurin series (after the English mathematician Colin Maclaurin (1698–1746), who published in 1742, but he too was anticipated in turn by James Stirling, who was mentioned earlier, just after (1.7.26)). What a tangled web history weaves for math historians!

which can be written as

$$(3.4.11) \qquad 1 - x\cot(x) = \frac{1}{3}x^2 + \frac{1}{45}x^4 + \frac{2}{945}x^6 + \frac{1}{4,725}x^8 + \cdots.$$

We'll return to (3.4.11) in just a few more steps.

But for now, look back at (3.3.14) and set $t = \pi$ in it to get

$$\cos(\alpha\pi) = \frac{\sin(\alpha\pi)}{\pi}\left[\frac{1}{\alpha} + 2\alpha\sum_{k=1}^{\infty}\frac{(-1)^k}{\alpha^2 - k^2}\cos(k\pi)\right]$$

or, since $\cos(k\pi) = (-1)^k$, and since $(-1)^k(-1)^k = (-1)^{2k} = 1$, we have

$$\frac{\cos(\alpha\pi)}{\sin(\alpha\pi)} = \cot(\alpha\pi) = \frac{1}{\pi}\left[\frac{1}{\alpha} + 2\alpha\sum_{k=1}^{\infty}\frac{1}{\alpha^2 - k^2}\right] = \frac{1}{\alpha\pi} + \frac{2\alpha\pi}{\pi^2}\sum_{k=1}^{\infty}\frac{1}{\alpha^2 - k^2}.$$

So,

$$\cot(\alpha\pi) = \frac{1}{\alpha\pi} + \sum_{k=1}^{\infty}\frac{2\alpha\pi}{(\alpha\pi)^2 - k^2\pi^2}$$

and, if we make the obvious change of variable to $x = \alpha\pi$, we have

$$\cot(x) = \frac{1}{x} + \sum_{k=1}^{\infty}\frac{2x}{x^2 - k^2\pi^2}$$

and this gives us

$$1 - x\cot(x) = -\sum_{k=1}^{\infty}\frac{2x^2}{x^2 - k^2\pi^2}$$

or

$$(3.4.12) \qquad 1 - x\cot(x) = \sum_{k=1}^{\infty} \frac{2x^2}{k^2\pi^2 - x^2}.$$

We can manipulate (3.4.12) just a bit more, as follows.

$$1 - x\cot(x) = \frac{2x^2}{\pi^2} \sum_{k=1}^{\infty} \frac{1}{k^2\left(1 - \frac{x^2}{k^2\pi^2}\right)}$$

$$= \frac{2x^2}{\pi^2} \sum_{k=1}^{\infty} \frac{1}{k^2}\left\{1 + \frac{x^2}{k^2\pi^2} + \frac{x^4}{k^4\pi^4} + \frac{x^6}{k^6\pi^6} + \cdots\right\},$$

which gives us

$$(3.4.13) \qquad 1 - x\cot(x) = x^2 \frac{2}{\pi^2} \sum_{k=1}^{\infty} \frac{1}{k^2} + x^4 \frac{2}{\pi^4} \sum_{k=1}^{\infty} \frac{1}{k^4}$$

$$+ x^6 \frac{2}{\pi^6} \sum_{k=1}^{\infty} \frac{1}{k^6} + x^8 \frac{2}{\pi^8} \sum_{k=1}^{\infty} \frac{1}{k^8} + \cdots.$$

Our final step is now another easy one: We simply equate the coefficients of equal powers of x in (3.4.13) and (3.4.11). This gives

$$\frac{2}{\pi^2} \sum_{k=1}^{\infty} \frac{1}{k^2} = \frac{1}{3} \text{ or } \sum_{k=1}^{\infty} \frac{1}{k^2} = \zeta(2) = \frac{\pi^2}{6}$$

$$\frac{2}{\pi^4} \sum_{k=1}^{\infty} \frac{1}{k^4} = \frac{1}{45} \text{ or } \sum_{k=1}^{\infty} \frac{1}{k^4} = \zeta(4) = \frac{\pi^4}{90}$$

$$\frac{2}{\pi^6} \sum_{k=1}^{\infty} \frac{1}{k^6} = \frac{2}{945} \text{ or } \sum_{k=1}^{\infty} \frac{1}{k^6} = \zeta(6) = \frac{\pi^6}{945}$$

$$\frac{2}{\pi^8}\sum_{k=1}^{\infty}\frac{1}{k^8}=\frac{1}{4,725} \text{ or } \sum_{k=1}^{\infty}\frac{1}{k^8}=\zeta(8)=\frac{\pi^8}{9,450}$$

and so on.

The use of Fourier series has reduced the calculation of $\zeta(2n)$ to a *cookbook algorithmic procedure,* one that takes a problem once considered to be profoundly mysterious and turns it into a routine homework exercise in freshman calculus. So, is this the end of the book? Well, of course, you know the answer is *no,* and that's because this use of Fourier series, alas, does not work for any of the $\zeta(2n + 1)$, starting with $\zeta(3)$, as I'll show you in the next section.

But before we do that, let me show you one more beautiful result that we can tease out of (3.4.13), a result that shows how all the zeta functions with positive, even integer arguments are intimately tied together. We start with a minor rewrite of (3.4.13), to define the function $p(x)$:

$$(3.4.14) \qquad \frac{1}{2}-\frac{x}{2}\cot(x)=\sum_{k=1}^{\infty}\left(\frac{x}{\pi}\right)^{2k}\zeta(2k)=p(x).$$

When you read of this approach in technical papers, the usual tale is that some unnamed person in the past one day noticed that $p(x)$ satisfies the differential equation[11]

$$(3.4.15) \qquad p^2(x)-\frac{1}{2}\frac{d}{dx}\{xp(x)\}=-\frac{1}{4}x^2,$$

and if you are puzzled by the use of the word *noticed,* you should realize that is how math journals encourage authors to save printer's ink and expensive page space (it's left up to the reader to fill in all the missing steps). Here's what you do next, where I'll ignore the issue

11. Plugging $p(x)=\frac{1}{2}-\frac{x}{2}\cot(x)$ into the left-hand side of (3.4.15) is a routine exercise in AP-calculus. You should, with little difficulty, be able to verify the claim that the result is, indeed, $-\frac{1}{4}x^2$.

of how in the world anyone would just *notice* the differential equation of (3.4.15).

Using the power series form of $p(x)$ in (3.4.15), we have

$$\left\{ \sum_{k=1}^{\infty} \left(\frac{x}{\pi} \right)^{2k} \zeta(2k) \right\}^2 - \frac{\pi}{2} \frac{d}{dx} \left\{ \sum_{j=1}^{\infty} \left(\frac{x}{\pi} \right)^{2j+1} \zeta(2j) \right\}^2 = -\frac{1}{4} x^2$$

or

$$\left\{ \sum_{k=1}^{\infty} \left(\frac{x}{\pi} \right)^{2k} \zeta(2k) \right\} \left\{ \sum_{n=1}^{\infty} \left(\frac{x}{\pi} \right)^{2n} \zeta(2n) \right\}$$

$$- \frac{\pi}{2} \sum_{j=1}^{\infty} \frac{(2j+1)}{\pi^{2j+1}} x^{2j} \zeta(2j) = -\frac{1}{4} x^2$$

or

$$(3.4.16) \qquad \left\{ \sum_{k=1}^{\infty} \left(\frac{x}{\pi} \right)^{2k} \zeta(2k) \right\} \left\{ \sum_{n=1}^{\infty} \left(\frac{x}{\pi} \right)^{2n} \zeta(2n) \right\}$$

$$- \sum_{j=1}^{\infty} \frac{2j+1}{2} \left(\frac{x}{\pi} \right)^{2j} \zeta(2j) = -\frac{1}{4} x^2.$$

There is a lot of information packed in (3.4.16).

Since (3.4.16) has to hold for all valid values of x, then we know the coefficients of the individual powers of x on the left-hand side must equal the coefficients of the same powers of x on the right-hand side. With that single observation, we can conclude the following. First, the two sums in curly brackets on the left each start with an x^2 term, and that means their product starts with an x^4 term. So, the only way to get an x^2 term to correspond with the x^2 term on the right is from the lone $j = 1$ term of the third sum on the left. That is, it must be true that

$$-\frac{2j+1}{2}\left(\frac{x}{\pi}\right)^{2j}\zeta(2j)\bigg|_{j=1}=-\frac{1}{4}x^2$$

or

$$\frac{3}{2}\frac{\zeta(2)}{\pi^2}=\frac{1}{4}$$

and so, at last,

$$\zeta(2)=\frac{2\pi^2}{12}=\frac{\pi^2}{6}$$

and we are off to a good start, as we have a result that is consistent with earlier calculations. We get something new, however, when we turn our attention to the terms beyond x^2.

Our second observation is that since there are no terms beyond x^2 on the right-hand side of (3.4.16), then the coefficients of the terms on the left-hand side, in powers of x^4 and higher, must all be equal to zero. So, suppose $j = t$, some fixed integer (equal to or greater than 2 but otherwise arbitrary) in the third sum on the left of (3.4.16). That gives us a term in x^{2t}. To cancel that term, we need to add all the terms produced by the product of the first two sums such that $2k + 2n = 2t$. So, since $k + n = t$, then as k increases from 1, we must have n varying as $t - k$ and so (3.4.16) becomes

$$\sum_{k=1}^{t-1}\left(\frac{x}{\pi}\right)^{2k}\zeta(2k)\left[\left(\frac{x}{\pi}\right)^{2(t-k)}\zeta(2\{t-k\})\right]-\left(\frac{x}{\pi}\right)^{2t}\frac{2t+1}{2}\zeta(2t)=0,$$

where you'll notice that the index k goes only up to $t - 1$ (ask yourself what would happen if $k \geq t$).

So, making the obvious cancellations of the x's and the π's, we instantly have our result, the elegant identity

(3.4.17) $$\sum_{k=1}^{t-1}\zeta(2k)\zeta(2\{t-k\})=\frac{2t+1}{2}\zeta(2t),\ t\ge 2$$

with $\zeta(2)=\frac{\pi^2}{6}$. If $t=2$, for example, (3.4.17) says $k=1$ (only), and we have

$$\zeta(2)\zeta(2)=\frac{5}{2}\zeta(4)$$

or

$$\zeta(4)=\frac{2}{5}\zeta^2(2)=\frac{2}{5}\left(\frac{\pi^2}{6}\right)^2=\frac{2\pi^4}{180}=\frac{\pi^4}{90},$$

which, again, agrees with a known result. The identity of (3.4.17) shows that the value of the zeta function with any positive, even integer argument is determined by *all* the values of the zeta function with even arguments that come before (see Challenge Problem 3.4.2).

Challenge Problem 3.4.1: Starting with (3.4.10), show that $1-\frac{1}{2^4}+\frac{1}{3^4}-\frac{1}{4^4}+\cdots=\frac{7\pi^4}{720}$.
Hint: Use (2.1.9) and (2.1.10).

Challenge Problem 3.4.2: Use (3.4.17) to find the exact expression for $\zeta(10)$.

3.5 How Fourier Series Fail to Compute $\zeta(3)$

In an attempt to find $\zeta(3)$ in the same way we found $\zeta(2)$ and $\zeta(4)$, we start by trying to simply mimic what we did in (3.4.1). That is, by writing

(3.5.1) $$f(t)=t^3,\ -\pi\le t\le\pi.$$

The periodic extension of this *odd* $f(t)$ over the entire t-axis will then involve only sine terms. Thus, with $T = 2\pi$ (and so $\omega_0 = 1$), we have $a_n = 0$ for all n, and so

$$f(t) = \sum_{n=1}^{\infty} b_n \sin(nt)$$

where

$$b_n = \frac{2}{T}\int_{-T/2}^{T/2} f(t)\sin(nt)dt = \frac{1}{\pi}\int_{-\pi}^{\pi} t^3 \sin(nt)dt$$

$$= \frac{1}{\pi}\left\{\left(\frac{3t^2}{n^2} - \frac{6}{n^4}\right)\sin(nt) + \left(\frac{6t}{n^3} - \frac{t^3}{n}\right)\cos(nt)\right\}\Big|_{-\pi}^{\pi}$$

$$= \frac{1}{\pi}\left\{\frac{12\pi}{n^3} - \frac{2\pi^3}{n}\right\}\cos(n\pi)$$

$$= \left\{\frac{12}{n^3} - \frac{2\pi^2}{n}\right\}\cos(n\pi)$$

and we arrive at

$$(3.5.2) \quad t^3 = \sum_{n=1}^{\infty}\left(\frac{12}{n^3} - \frac{2\pi^2}{n}\right)\cos(n\pi)\sin(nt), \quad -\pi \le t \le \pi.$$

If we set $t = \pi$ in (3.5.2), we get $\pi^3 = 0$, a dubious claim that is clearly not going to be of much help. You can confirm for yourself that there is, in fact, no value to which we can set t equal to in (3.5.2) that will give us $\zeta(3)$. What happened? Suddenly the Fourier method that worked so well in the previous section for $\zeta(2)$ and $\zeta(4)$—and, indeed, for all the $\zeta(2n)$—has failed us for $\zeta(3)$.

You might wonder if the immediate explanation for this failure is that the Fourier series expansion of a discontinuous function converges, at a discontinuity, to the average of the function values on each side of the discontinuity. Our odd, periodically extended t^3 is discontinuous at $t = \pi$ (just like the periodic extension of the odd function t, shown in Figure 3.2.2), and you'll notice that the average of π^3 and $-\pi^3$ is zero, which is indeed the value to which (3.5.2) converges at $t = \pi$. Well then, you might go on to argue, the extension of an odd function is not a precise mimicking of what we did for the even t^2 and t^4; so, what if we make a periodic extension of t^3 that is *even* (and so continuous) over the interval $-\pi \leq t \leq \pi$? That is, let's calculate the Fourier series of the periodic extension of

$$(3.5.3) \qquad\qquad f(t) = \begin{cases} t^3, 0 \leq t \leq \pi \\ -t^3, -\pi \leq t \leq 0 \end{cases}.$$

That should give us a Fourier series that does converge to π^3 at $t = \pi$. Let's try that.[12]

We still have $T = 2\pi$ (and $\omega_0 = 1$), but now the Fourier series will have only cosine terms ($b_n = 0$ for all n). That is,

$$f(t) = \frac{1}{2}a_0 + \sum_{n=1}^{\infty} a_n \cos(nt),$$

where

$$a_0 = \frac{1}{\pi}\int_{-\pi}^{0} -t^3 dt + \frac{1}{\pi}\int_{0}^{\pi} t^3 dt = \frac{1}{\pi}\left[\left(-\frac{1}{4}t^4\right)\Big|_{-\pi}^{0} + \left(\frac{1}{4}t^4\right)\Big|_{0}^{\pi}\right]$$

$$= \frac{\pi^4}{4\pi} + \frac{\pi^4}{4\pi} = \frac{\pi^3}{2}.$$

12. What we are doing here is called, by physicists and engineers, *experimenting* (or, even more bluntly, *messing around to see what happens*), and despite the sanitized math you find in textbooks and published journal papers, this is what mathematicians do, too, in the privacy of their offices when struggling with a new problem.

Also,

$$a_n = \frac{2}{T}\int_{-\frac{T}{2}}^{\frac{T}{2}} f(t)\cos(nt)\,dt = \frac{1}{\pi}\left[\int_{-\pi}^{0} -t^3\cos(nt)\,dt + \int_{0}^{\pi} t^3\cos(nt)\,dt\right].$$

In the first integral in the square brackets, let $u = -t$ $(du = -dt)$ and so

$$a_n = \frac{1}{\pi}\left[\int_{\pi}^{0} u^3\cos(-nu)(-du) + \int_{0}^{\pi} t^3\cos(nt)\,dt\right]$$

$$= \frac{1}{\pi}\left[\int_{0}^{\pi} u^3\cos(nu)\,du + \int_{0}^{\pi} t^3\cos(nt)\,dt\right] = \frac{2}{\pi}\int_{0}^{\pi} t^3\cos(nt)\,dt$$

$$= \frac{2}{\pi}\left\{\left(\frac{3t^2}{n^2} - \frac{6}{n^4}\right)\cos(nt) + \left(\frac{t^3}{n} - \frac{6t}{n^3}\right)\sin(nt)\right\}\Big|_{0}^{\pi}$$

$$= \frac{2}{\pi}\left\{\left(\frac{3\pi^2}{n^2} - \frac{6}{n^4}\right)\cos(n\pi) + \frac{6}{n^4}\right\}$$

and we see we are dead in the water right here, as there is no $1/n^3$ term. That is, $\zeta(3)$ cannot possibly appear in the Fourier series for $f(t)$ for any value of t.

Well, okay, we have to admit that this new attempt didn't work so well, either. So let's try messing around with something really different, this time with the Fourier series of an unbounded function. Maybe doing something that dramatically "off the wall" will accomplish the job of calculating $\zeta(3)$. (Hope springs eternal!) If you look back at (2.4.22), you'll see there that we (Euler) derived the expression

$$\ln\{\sin(x)\} = -\sum_{n=1}^{\infty} \frac{\cos(2nx)}{n} - \ln(2)$$

or, if we write $x = \frac{t}{2}$,

$$\ln\left\{\sin\left(\frac{t}{2}\right)\right\} + \ln(2) = \ln\left\{2\sin\left(\frac{t}{2}\right)\right\} = -\sum_{n=1}^{\infty} \frac{\cos(nt)}{n}.$$

That is,

$$(3.5.4) \qquad \sum_{n=1}^{\infty} \frac{\cos(nt)}{n} = -\ln\left\{2\sin\left(\frac{t}{2}\right)\right\}.$$

If you look closely at (3.5.4), you'll recognize that the left-hand side is a Fourier series,[13] and so we see that we have almost instantly gotten our hands on the Fourier series for $-\ln\{2\sin(\frac{t}{2})\}$ without performing the usual computations. (We should actually write $-\ln\{|2\sin(\frac{t}{2})|\}$, where the absolute value signs prevent the argument of the log function from being negative, which would make the log function imaginary.) The period of the Fourier series on the left of (3.5.4) is clearly 2π, as shown in Figure 3.5.1, which also illustrates that we are dealing with an unbounded function (both sides of (3.5.4) obviously blow up when t is any integer multiple of 2π).

This result may remind you of Euler's trigonometric series in (2.3.1) for $\frac{\pi-t}{2}$, in which all the cosines of (3.5.4) are replaced with sines. But, of course, what a difference it makes to simply shift each of Euler's sines into a cosine! The right-hand side of (3.5.4) does

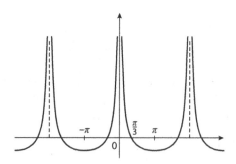

FIGURE 3.5.1.

The periodic, unbounded function $-\ln|2\sin(\frac{t}{2})|, -\infty < t < \infty$, where t is the horizontal axis.

13. Another example of Euler encountering a Fourier series *before* Fourier!

indeed blow up if t is any integer multiple of 2π, as the trigonometric series becomes the harmonic series. The blow-up is a very slow one, of course, as you'll recall that the harmonic series diverges only as the logarithm of the number of terms.

To get something involving $\zeta(3)$ out of (3.5.4), let's integrate (3.5.4) over the interval

$$0 \le t \le y:$$

$$-\int_0^y \ln\left\{2\sin\left(\frac{t}{2}\right)\right\} dt = \left(\sum_{n=1}^\infty \frac{\sin(nt)}{n^2}\right)\Big|_0^y = \sum_{n=1}^\infty \frac{\sin(ny)}{n^2}.$$

Then, integrating again, now over the interval $0 \le y \le u$,

$$-\int_0^u \left(\int_0^y \ln\left\{2\sin\left(\frac{t}{2}\right)\right\} dt\right) dy = \left(-\sum_{n=1}^\infty \frac{\cos(ny)}{n^3}\right)\Big|_0^u$$

$$= -\sum_{n=1}^\infty \frac{\cos(nu)}{n^3} + \sum_{n=1}^\infty \frac{1}{n^3}$$

$$= -\sum_{n=1}^\infty \frac{\cos(nu)}{n^3} + \zeta(3)$$

or

$$(3.5.5) \quad \zeta(3) = \sum_{n=1}^\infty \frac{\cos(nu)}{n^3} - \int_0^u \left(\int_0^y \ln\left\{2\sin\left(\frac{t}{2}\right)\right\} dt\right) dy.$$

Alas, there are no values for u and y such that (3.5.5) can be evaluated to give us $\zeta(3)$. Of course, setting $u = 0$ results in the double integral vanishing, and that gives us $\zeta(3) = \sum_{n=1}^\infty \frac{1}{n^3}$, but that's hardly breaking news. You knew that before you got out of bed this morning!

With our continuing failures here (and dozens more other flops that mathematicians have suffered over the past 250 years), it's hard to escape the feeling that God, when He made the World, intentionally (for whatever reason) threw a tightly buttoned cloak over the values of all the $\zeta(2n + 1)$, starting with $\zeta(3)$. The zeta-3 problem is, for mathematicians, like the massive rock the mythologically cursed Sisyphus is forever almost pushing up to the top of a hill, only to have it *every single time* slip from his grasp at the very last moment before success and so have to helplessly watch it roll back down to the bottom of the hill. A feeling of increasing despair that rivals Sisyphus' has haunted mathematicians through the now nearly three centuries since Euler calculated $\zeta(2n)$, and one can hear their quiet dread in these words, written after yet a different attack than I've done here also failed to pull aside God's Cloak: "but *as usual* [my emphasis] the value of $\Sigma(1/n^3)$ remains unrevealed."[14]

Challenge Problem 3.5.1: Back in Section 3.3, I showed you how the Gregory/Leibniz series $\frac{\pi}{4} = 1 - \frac{1}{3} + \frac{1}{5} - \frac{1}{7} + \cdots$ comes out of (2.3.1) if we substitute $t = \frac{\pi}{2}$ into Euler's trigonometric series. A much less well-known series that can also be generated from (2.3.1) is $\frac{\pi\sqrt{2}}{4} = 1 + \frac{1}{3} - \frac{1}{5} - \frac{1}{7} + \frac{1}{9} + \frac{1}{11} - - + + \cdots$, which looks "sorta like" its better-known cousin but with signs that alternate every two terms. (Euler attributed this series to Newton.) To convince you of the likely truth of this claim, note that $\frac{\pi\sqrt{2}}{4} = 1.1107207 \ldots$, while the sum of the first 1 million terms of the series is $1.1107202 \ldots$. Explain how to derive this series. Hint: Try setting $t = \frac{\pi}{4}$ in (2.3.1) and use the Gregory/Leibniz series at the appropriate place in your analysis.

3.6 Fourier Transforms and Poisson Summation

In this section we'll use Fourier theory to derive a result that will be central to the development (in the next section) of the functional equation of the zeta function. This preliminary result, called

---

14. Ralph Palmer Agnew, *Differential Equations* (McGraw-Hill, 1960), p. 367. But see the hopeful words of the Comte de Buffon at the end of the Epilogue.

Poisson's summation formula, is not without some irony, as Poisson (mentioned in Challenge Problem 2.1.2) was one of the more severe critics of Fourier's mathematics. To derive this preliminary result, however, requires that we first derive a preliminary result for *it*. A second-order preliminary, if you will!

Fourier series are, as we've been extensively discussing, the mathematical description of a periodic function $f(t)$—but what if we have a function $f(t)$, defined over $-\infty < t < \infty$, that is not periodic? Since such a function "uses up" the entire infinite t-axis, we can't employ our earlier trick of periodically extending $f(t)$ to make a periodic function, because there is nowhere left on the t-axis to extend $f(t)$ into. So, a Fourier *series* approach is out. But that doesn't mean we have hit a brick wall, as there is yet a new trick we can unleash (the clever mind of Fourier was not easily stumped).

Here's what we are going to do. Think of our non-periodic $f(t)$ as being periodic with an infinite period, and so it "really is" periodic, and we are simply observing, as t goes from minus infinity to plus infinity, the period we just happen to be living through. Is this a fiendishly clever idea, perhaps even an idea with a deep metaphysical significance[15] that mere mathematical minds might pursue with some caution? **YES!** Actually, it's a full-blown outrageous idea (and so you can be sure that Euler would have instantly embraced it) and, in the spirit of this book, one that we also can't resist.

Mathematically, this idea takes the complex Fourier series equations of (3.3.1) and (3.3.5), which I'll rewrite here as

$$(3.6.1) \qquad f(t) = \sum_{k=-\infty}^{\infty} c_k e^{ik\omega_0 t}, \ \omega_0 T = 2\pi$$

---

15. What was going on in the period before the "current" period? What will happen in the "next" period? Such questions, while interesting, aren't mathematics, and I'll leave them for the philosophers.

and

$$(3.6.2) \qquad c_k = \frac{1}{T}\int_{period} f(t)e^{-ik\omega_0 t}\,dt,\ \omega_0 T = 2\pi$$

and explores what happens to them as $T \to \infty$. Perhaps *explore* is too gentle a word—what I'll do is play with these two Fourier series equations in a pretty casual, formal way (remember how we pushed symbols with unbounded enthusiasm back in Chapter 1 to get Ramanujan's Master Theorem?), with little (if any) regard to justifying the manipulations. But—and this is most important to appreciate—once we are done with all the symbol pushing, *none of our superficial sloppiness will matter*. Once we have the "mathematical answer" to our immediate question (what happens as $T \to \infty$?), we can forget how we got that "answer" and simply treat it as a definition. Indeed, there are books written by pure mathematicians, in discussions of the $T \to \infty$ issue, that take precisely the same approach that I'll now show you, and so we are not in complete disgrace.

To start, make the obvious observation that $k\omega_0$ in the exponent of the c_k integral in (3.6.2) changes by ω_0 as k increases from one integer value to the next. If we call this change $\Delta\omega$, then $\Delta\omega = \omega_0$. Now, since $\omega_0 = \frac{2\pi}{T}$, then as $T \to \infty$, we see $\Delta\omega = \omega_0 \to 0$, and so we'll write $d\omega$ instead of $\Delta\omega$. That is, as $T \to \infty$, the change in frequency becomes a *differential* change. Thus, to summarize,

$$\lim_{T\to\infty} \Delta\omega = \lim_{T\to\infty} \omega_0 = \lim_{T\to\infty} \frac{2\pi}{T} = d\omega.$$

In addition, it follows that, as $T \to \infty$, $k\left(\frac{2\pi}{T}\right) \to k\,d\omega$. Since $d\omega$ is infinitesimally small, as k varies from $-\infty$ to $+\infty$, we should see $k\,d\omega$ behave like a continuous variable, that is, like ω. So,

$$\lim_{T\to\infty} k\omega_0 = \omega.$$

If you accept all this (or at least are willing to hang around for a while to see where this is going), we can rewrite (3.6.2), as $T \to \infty$ (with the period symmetrical around $t = 0$), as

$$\lim_{T \to \infty} c_k = \lim_{T \to \infty} \frac{1}{T} \int_{-T/2}^{T/2} f(t) e^{-ik\omega_0 t} dt = \lim_{T \to \infty} \frac{1}{2\pi} \frac{2\pi}{T} \int_{-T/2}^{T/2} f(t) e^{-ik\omega_0 t} dt$$

$$= \lim_{T \to \infty} \frac{1}{2\pi} \left[\int_{-T/2}^{T/2} f(t) e^{-ik\omega_0 t} dt \right] \frac{2\pi}{T}$$

or

$$\lim_{T \to \infty} c_k = \frac{1}{2\pi} \left[\int_{-\infty}^{\infty} f(t) e^{-ik\omega_0 t} dt \right] d\omega.$$

Or, if we define the last integral in the square brackets as the so-called *Fourier transform* of $f(t)$, written as $F(\omega)$, then we have

(3.6.3) $$\lim_{T \to \infty} c_k = \frac{1}{2\pi} F(\omega) d\omega$$

where

(3.6.4) $$F(\omega) = \int_{-\infty}^{\infty} f(t) e^{-i\omega t} dt.$$

Finally, to close the loop, let's now insert (3.6.3) back into (3.6.1) to get

$$f(t) = \sum_{k=-\infty}^{\infty} \left\{ \frac{1}{2\pi} F(\omega) d\omega \right\} e^{ik\omega_0 t} = \frac{1}{2\pi} \sum_{k=-\infty}^{\infty} F(\omega) e^{ik\omega_0 t} d\omega.$$

Now, as it stands, this last expression is a mixed bag of notation, since parts of it are for $T \to \infty$ and other parts of it are written as if T were still finite. If we imagine $T \to \infty$ all through the last expression, then the summation becomes an integral and we arrive at

$$(3.6.5) \qquad f(t) = \frac{1}{2\pi} \int_{-\infty}^{\infty} F(\omega) e^{i\omega t} d\omega.$$

These last two equations, (3.6.4) and (3.6.5), form what is called the *Fourier transform pair*, written as $f(t) \leftrightarrow F(\omega)$, where the double-headed arrow means that each side completely determines the other (to support this claim of unique, one-to-one correspondence between a function and its Fourier transform requires a proof, which I'll let you pursue in a math book deeper than this one). This completes the second-order preliminary result I mentioned at the start. We are now ready for the first-order preliminary result that we'll use in the next section to derive the functional equation of the zeta function.

Suppose we have a function $f(t)$ defined over the entire infinite line $-\infty < t < \infty$. From this $f(t)$ we then construct another function, $g(t)$, defined as

$$(3.6.6) \qquad g(t) = \sum_{k=-\infty}^{\infty} f(t+k).$$

From (3.6.6) we have

$$g(t+1) = \sum_{k=-\infty}^{\infty} f(t+k+1),$$

which, with a change in the summation index to $n = k + 1$, becomes

$$(3.6.7) \qquad g(t+1) = \sum_{n=-\infty}^{\infty} f(t+n) = g(t).$$

That is, $g(t)$ is periodic with period $T = 1$.

Since $g(t)$ is periodic, it can be written as a Fourier series with $\omega_0 = 2\pi$ (since $T = 1$), and so from (3.6.1) and (3.6.2) we have

$$(3.6.8) \qquad g(t) = \sum_{n=-\infty}^{\infty} c_n e^{in2\pi t}$$

where

$$c_n = \frac{1}{T} \int_{period} g(t) e^{-in2\pi t} dt = \int_0^1 \sum_{k=-\infty}^{\infty} f(t+k) e^{-in2\pi t} dt$$

or

$$(3.6.9) \qquad c_n = \sum_{k=-\infty}^{\infty} \int_0^1 f(t+k) e^{-in2\pi t} dt.$$

If we change variable in (3.6.9) to $s = t + k$ (and so $ds = dt$), then

$$c_n = \sum_{k=-\infty}^{\infty} \int_k^{k+1} f(s) e^{-in2\pi(s-k)} ds = \sum_{k=-\infty}^{\infty} \int_k^{k+1} f(s) e^{-in2\pi s} e^{ink2\pi} ds$$

or, as n and k are both integers, then Euler's identity says $e^{ink2\pi} = 1$ and therefore

$$(3.6.10) \qquad c_n = \sum_{k=-\infty}^{\infty} \int_k^{k+1} f(s) e^{-in2\pi s} ds.$$

Finally, observing that

$$\sum_{k=-\infty}^{\infty} \int_k^{k+1} = \int_{-\infty}^{\infty},$$

we see that (3.6.10) becomes

$$(3.6.11) \qquad c_n = \int_{-\infty}^{\infty} f(s) e^{-in2\pi s} ds = \int_{-\infty}^{\infty} f(t) e^{-in2\pi t} dt,$$

where the dummy variable s has been replaced by the dummy variable t.

Now, look back at (3.6.4) and you'll see that (3.6.11) is

$$c_n = F(2\pi n).$$

Putting this into (3.6.8), and using (3.6.6), we arrive at

$$(3.6.12) \qquad g(t) = \sum_{n=-\infty}^{\infty} F(2\pi n)e^{in2\pi t} = \sum_{k=-\infty}^{\infty} f(t+k).$$

This is an identity in t and, for $t = 0$ in particular, (3.6.12) reduces to Poisson's summation formula, derived in 1827:

$$(3.6.13) \qquad \sum_{k=-\infty}^{\infty} f(k) = \sum_{n=-\infty}^{\infty} F(2\pi n),$$

which connects a sum (not an integral) over a function, to a sum (not an integral) over the Fourier transform of that function. The expression in (3.6.13) might look boringly benign but that's not so, as the penultimate calculation of this section (giving the result we'll need in the next section for the functional equation of the zeta function) will now show you.

We start with the so-called Gaussian pulse function,[16]

$$(3.6.14) \qquad f(t) = e^{-\alpha t^2}, \alpha > 0, -\infty < t < \infty,$$

which, from (3.6.4), has the Fourier transform

---

16. This function, because of its many nice mathematical properties, is a favorite of physicists and electrical engineers. That's because if we think of t as time, the Gaussian pulse is a continuous, endlessly differentiable description of a quantity that is localized in time (around time $t = 0$); for example, the transmission of a pulse of energy.

$$(3.6.15) \qquad F(\omega) = \int_{-\infty}^{\infty} e^{-\alpha t^2} e^{-i\omega t} dt.$$

Doing the integral in (3.6.15) may appear to be an intimidating task, but there is a clever, elementary way to do it. If we make the assumption that we can interchange the order of differentiation and integration (not always true, but it can be shown to be so here), then differentiating (3.6.15) with respect to ω gives

$$\frac{dF}{d\omega} = \frac{d}{d\omega} \int_{-\infty}^{\infty} e^{-\alpha t^2} e^{-i\omega t} dt = \int_{-\infty}^{\infty} e^{-\alpha t^2} \frac{d}{d\omega} (e^{-i\omega t}) dt = -i \int_{-\infty}^{\infty} t e^{-\alpha t^2} e^{-i\omega t} dt.$$

If we evaluate the last integral using integration by parts,[17] we get

$$\int_{-\infty}^{\infty} t e^{-\alpha t^2} e^{-i\omega t} dt = \left(-\frac{1}{2\alpha} e^{-\alpha t^2} e^{-i\omega t} \right) \Big|_{-\infty}^{\infty} - i \frac{\omega}{2\alpha} \int_{-\infty}^{\infty} e^{-\alpha t^2} e^{-i\omega t} dt$$

$$= -i \frac{\omega}{2\alpha} \int_{-\infty}^{\infty} e^{-\alpha t^2} e^{-i\omega t} dt,$$

because $\lim_{t \to \pm\infty} e^{-\alpha t^2} e^{-i\omega t} = 0$. But this last integral is $F(\omega)$, and so (remembering that $i^2 = -1$) we have a simple first-order differential equation for $F(\omega)$:

$$\frac{dF}{d\omega} = -i \left\{ -i \frac{\omega}{2\alpha} F(\omega) \right\} = -\frac{\omega}{2\alpha} F(\omega)$$

or, separating the variables F and ω,

$$(3.6.16) \qquad \frac{dF}{F} = -\frac{\omega}{2\alpha} d\omega.$$

---

17. In the classic formula from freshman calculus, $\int_{-\infty}^{\infty} u\,dv = (uv)\big|_{-\infty}^{\infty} - \int_{-\infty}^{\infty} v\,du$, let $u = e^{-i\omega t}$ and $dv = t e^{-\alpha t^2} dt$. Then $du = -i\omega e^{-i\omega t} dt$ and $v = -\frac{1}{2\alpha} e^{-\alpha t^2}$.

Writing $\ln(C)$ as the constant of indefinite integration, (3.6.16) integrates by inspection to

$$\ln\{F(\omega)\} = -\frac{\omega^2}{4\alpha} + \ln(C)$$

or

$$F(\omega) = Ce^{-\frac{\omega^2}{4\alpha}}.$$

To determine C, notice that $C = F(0)$, which from (3.6.15) says

$$C = \int_{-\infty}^{\infty} e^{-\alpha t^2} dt.$$

This integral is equal[18] to $\sqrt{\frac{\pi}{\alpha}}$, and so we have the Fourier transform pair

(3.6.17) $f(t) = e^{-\alpha t^2} \leftrightarrow F(\omega) = \sqrt{\frac{\pi}{\alpha}} e^{-\frac{\omega^2}{4\alpha}}.$

Notice that (3.6.17) says a Gaussian pulse in the time domain has a Fourier transform that is also a Gaussian pulse in the frequency domain.

Using (3.6.17) in the Poisson summation formula of (3.6.13) gives us our prize, the wonderful identity

(3.6.18) $\sum_{k=-\infty}^{\infty} e^{-\alpha k^2} = \sqrt{\frac{\pi}{\alpha}} \sum_{n=-\infty}^{\infty} e^{-\frac{\pi^2 n^2}{\alpha}},$

---

18. In Appendix 2, it is shown that $\int_0^{\infty} e^{-x^2} dx = \frac{1}{2}\sqrt{\pi}$. Since the integrand is even, we immediately have $\int_{-\infty}^{\infty} e^{-x^2} dx = \sqrt{\pi}$, and a change of variable to $x = t\sqrt{\alpha}$ gives $\int_{-\infty}^{\infty} e^{-\alpha t^2} dt = \sqrt{\frac{\pi}{\alpha}}$.

which we will find invaluable in the next section, where I'll show you how Riemann used (3.6.18) in a derivation of the functional equation of the zeta function.

Now, before we leave this section, let me end it by showing you a beautiful physical interpretation of the Fourier transform, just to convince you that all of our fancy math hasn't left the real world behind. You'll recall from (3.3.9) that we defined the energy of a real-valued, periodic $f(t)$ to be

$$W = \int_{period} f^2(t)dt.$$

(We talked in terms of energy *per period*, that is, power for a periodic function, because, obviously, the total energy over all time of any periodic, endlessly repeating function is infinite.) For a non-periodic, real-valued $f(t)$ defined over all time (as we are considering here), the total energy can be finite, and we simply extend the energy definition in the obvious way to

$$W = \int_{-\infty}^{\infty} f^2(t)dt.$$

We can write W in terms of the Fourier transform of $f(t)$ as follows. $F(\omega)$ as written in (3.6.4) is, in general, complex, and its complex conjugate is found (as with any complex quantity) by simply reversing the sign of every appearance of $i = \sqrt{-1}$ to give

$$F^*(\omega) = \int_{-\infty}^{\infty} f(t)e^{i\omega t}dt.$$

Now, writing $f^2(t) = f(t)f(t)$, we have

$$W = \int_{-\infty}^{\infty} f(t)f(t)dt = \int_{-\infty}^{\infty} f(t)\left\{\frac{1}{2\pi}\int_{-\infty}^{\infty} F(\omega)e^{i\omega t}d\omega\right\}dt$$

where (3.6.5) has been used to replace the second $f(t)$ in $f(t)f(t)$. If, as usual, we assume we can reverse the order of integration, then

$$W = \int_{-\infty}^{\infty} \frac{1}{2\pi} F(\omega) \left\{ \int_{-\infty}^{\infty} f(t) e^{i\omega t} dt \right\} d\omega$$

$$= \int_{-\infty}^{\infty} \frac{1}{2\pi} F(\omega) F^*(\omega) d\omega = \int_{-\infty}^{\infty} \frac{1}{2\pi} |F(\omega)|^2 \, d\omega.$$

It is easy to show that $|F(\omega)|^2$ is an even function[19] of ω, and so

$$W = 2 \int_0^{\infty} \frac{1}{2\pi} |F(\omega)|^2 \, d\omega = \int_0^{\infty} \frac{1}{\pi} |F(\omega)|^2 \, d\omega.$$

The quantity $\frac{1}{\pi} |F(\omega)|^2$ is called the *energy spectral density* of $f(t)$. The energy spectral density represents how the energy of $f(t)$ is distributed over frequency, a result called Rayleigh's energy theorem, after the English mathematical physicist John William Strutt (1842–1919), better known as Lord Rayleigh (winner of the 1904 Nobel Prize in physics), who stated it in 1889. Rayleigh's energy theorem is the non-periodic version of Parseval's power formula in (3.3.11) for infinite energy, periodic functions. In general,

$$(3.6.19) \qquad\qquad \int_{\omega_1}^{\omega_2} \frac{1}{\pi} |F(\omega)|^2 \, d\omega$$

is the energy of $f(t)$ that is in the frequency interval $\omega_1 < \omega < \omega_2$.

Challenge Problem 3.6.1: The identity in (3.6.18) *is* pretty amazing and, since we got it through what many might charitably call a "maze of manipulations," you might be secretly wondering whether (3.6.18) is right. So, (a) confirm that if $a = \pi$, then (3.6.18) immediately reduces to an obviously true identity. (b) As a more detailed check for the general case of $a \neq \pi$, suppose we pick (for no special reason) $a = 1$. Then (3.6.18) becomes the claim $\sum_{k=-\infty}^{\infty} e^{-k^2} = \sqrt{\pi} \sum_{n=-\infty}^{\infty} e^{-\pi^2 n^2}$ or

---

19. Since $F(\omega) = \int_{-\infty}^{\infty} f(t) e^{-i\omega t} dt$, then $F(-\omega) = \int_{-\infty}^{\infty} f(t) e^{i\omega t} dt = F^*(\omega)$. Now, $|F(\omega)|^2 = F(\omega) F^*(\omega) = F(\omega) F(-\omega)$, which is obviously unchanged if we write $-\omega$ for each ω. That is, $|F(\omega)|^2$ is even.

(as you should confirm) $1+2\Sigma_{k=1}^{\infty}e^{-k^2} = \sqrt{\pi}(1+2\Sigma_{n=1}^{\infty}e^{-\pi^2 n^2})$. Since the exponentials go to zero pretty fast as k and n increase, the numerical values of each side of this last expression should be closely approximated with the use of only the first few terms. In fact, by direct numerical calculation, show that using just the first three (!) terms in each sum gives numbers that don't begin to disagree until the sixth decimal place.

Challenge Problem 3.6.2: If you look in any good book of math tables, in the section on definite integrals, you'll find the entry $\int_0^{\infty}\frac{\sin^2(ax)}{x^2}dx = \frac{\pi a}{2}$. This important integral occurs in numerous applications in physics (optics) and electrical engineering (communication and information theory). See if you can derive it. Hint: Start with $f(t) = \{_{0,\text{ otherwise}}^{1, 0 < t < 1}$, and then use Rayleigh's energy theorem.

Challenge Problem 3.6.3: In note 14 of Chapter 1, I told you about Dirac's impulse function $\delta(t) = \{_{0, t \neq 0}^{\infty, t=0}$, with the property $\int_{-\infty}^{\infty}\delta(t)dt = 1$. An interesting question is: What is the energy of $\delta(t)$? One might be tempted to say that it's infinite, because at $t = 0$ the function is infinite. But one might be equally tempted to reject that, because $\delta(t)$ is non-zero only for a time interval of zero duration. You now know enough to resolve this puzzle, as outlined in the following three steps: (a) If $\varphi(t)$ is a smoothly varying, always finite (but otherwise arbitrary) function, give a "justification" for why $\int_{-\infty}^{\infty}\delta(t)\varphi(t)dt = \varphi(0)$. Hint: Don't be afraid to think like a physicist or an engineer (in other words, two plus two is still four but otherwise be fearless). (b) Calculate the Fourier transform of $\delta(t)$, using the result of (a). Hint: Let $\varphi(t) = e^{-i\omega t}$. (c) Calculate the energy of $\delta(t)$, using the result of (b) and Rayleigh's energy theorem.

3.7 The Functional Equation of the Zeta Function

This final section of the chapter is all about the genius of Bernhard Riemann. I touched on him just a bit at the start of Chapter 1, and here I'll elaborate on what I wrote there. He died far too young, of tuberculosis, at age 39, and yet though he was just reaching the full power of his intellect when he departed this life, he left mathematics with what professional mathematicians think is its greatest unsolved problem, a problem that has often been described as the ultimate Holy Grail of mathematics. It's a problem *so* difficult, and *so* mysterious, that many mathematicians have seriously entertained the possibility that it can't be solved.

Well, you might ask, why isn't this book about *that* problem, instead of being about zeta-3? The answer is that Riemann's problem doesn't satisfy all my selection criteria; in particular, a grammar school student is going to need a *lot* more than he/she has got to understand the question. So, the first part of this section is my attempt to fill in that missing background. The zeta-3 problem is, however, closely linked to Riemann's problem, because both involve the zeta function.

Riemann's problem is the famous Riemann Hypothesis (RH), discussed in the second box of Section 1.2, a conjecture which has so far soundly defeated (since Riemann formulated it in 1859) all the efforts of the greatest mathematical minds in the world (including Riemann's) to either prove or disprove it. Forty years after its conjecture, and with no solution in sight, the great German mathematician David Hilbert (1862–1943) decided to add some incentive. In 1900, at the Second International Congress of Mathematicians in Paris, he gave a famous talk titled "Mathematical Problems." During that talk he discussed some problems that he felt represented potentially fruitful directions for future research. The problems included, for example, determining the transcendental nature (or not) of $2^{\sqrt{2}}$, resolving the issue of Fermat's Last Theorem (FLT), and resolving the RH, in *decreasing* order of difficulty (in Hilbert's estimation).

All of Hilbert's problems became famous overnight, and to solve one brought instant celebrity among fellow mathematicians. Hilbert's own estimate of the difficulty of his problems was slightly askew, however, as the $2^{\sqrt{2}}$ issue was settled by 1930 (it *is* transcendental), and Fermat's Last Theorem was laid to rest by the mid-1990s. The RH, however, the presumed easiest of the three, has proven itself to be the toughest. Hilbert eventually came to appreciate this. A well-known story in mathematical lore says he once remarked that, if he awoke after sleeping for 500 years, the first question he would ask is: "Has the Riemann hypothesis been proven?" The answer is currently still *no*. A century after Hilbert's famous talk in Paris, the Clay Mathematics Institute in Cambridge, MA, proposed in 2000 seven so-called "Millennium Prize Problems," with each to be worth a 1 million dollar award to its solver. The RH is one of those elite problems and, as I write in 2021, the 1 million dollars for its solution remains unclaimed.

The RH is important for more than just being famous for being unsolved; there are numerous theorems in mathematics, all of which mathematicians believe to be correct, that are based on the *assumed* truth of the RH. If the RH is someday shown to be false, the existing proofs of all those theorems collapse, and they will have to be revisited and new proofs (hopefully) found.

Riemann began his study of the zeta function

$$\zeta(s) = \sum_{n=1}^{\infty} \frac{1}{n^s} = 1 + \frac{1}{2^s} + \frac{1}{3^s} + \cdots$$

because of Euler's connection of it to the primes in (1.3.11) (Riemann called it his "point of departure"), with the thought that studying $\zeta(s)$ would aid in his quest for a formula for $\pi(x)$, defined to be the number of primes not greater than x. $\pi(x)$ is a measure of how the primes are distributed among the integers. It should be obvious to you that $\pi(1/2) = 0$, that $\pi(2) = 1$, and that $\pi(6) = 3$, but perhaps it is not quite so obvious that $\pi(10^{18}) = 24{,}739{,}954{,}287{,}740{,}860$. When Riemann started his studies of the distribution of the primes, one known approximation to $\pi(x)$ was the so-called logarithmic integral, written as

$$\mathrm{li}(x) = \int_2^x \frac{du}{\ln(u)},$$

which is actually a pretty good approximation to $\pi(x)$. For example,

$$\frac{\pi(1,000)}{\text{li}(1,000)} = \frac{168}{178} = 0.94\ldots,$$

$$\frac{\pi(100,000)}{\text{li}(100,000)} = \frac{9,592}{9,630} = 0.99\ldots,$$

$$\frac{\pi(100,000,000)}{\text{li}(100,000,000)} = \frac{5,761,455}{5,762,209} = 0.999\ldots,$$

$$\frac{\pi(1,000,000,000)}{\text{li}(1,000,000,000)} = \frac{50,847,478}{50,849,235} = 0.9999\ldots.$$

In an 1849 letter, Gauss, who signed off on Riemann's 1851 doctoral dissertation with a glowing endorsement, claimed to have known of this behavior of li(x) since 1791 or 1792, when he was just 14. With what is known of Gauss' genius, there is little doubt that is true.

Numerical calculations like those above immediately suggest the conjecture

$$\lim_{x\to\infty} \frac{\pi(x)}{\text{li}(x)} = 1,$$

which is a statement of what mathematicians call the prime number theorem. Although highly suggestive, such numerical calculations of course prove nothing, and in fact it wasn't until 1896 that mathematical proofs of the prime number theorem were simultaneously and independently discovered by Charles-Joseph de la Vallée-Poussin (1866–1962) in Belgium and Jacques Hadamard (1865–1963) in France. Each man used very advanced techniques from complex variable function theory, applied to the zeta function. It was a similar quest (using the zeta function as well) that Riemann was on in 1859, years before either Vallée-Poussin or Hadamard had been born.

To start his work, Riemann immediately tackled a technical issue concerning the very definition of $\zeta(s)$, namely, the sum converges only if $s > 1$. More generally, if we extend the s in Euler's definition of the zeta function in (1.2.6) from being real to being complex (that is, if we write $s = \sigma + it$), then $\zeta(s)$ makes sense only if $\sigma > 1$. Riemann, however, wanted to be able to treat $\zeta(s)$ as defined everywhere in the complex plane or, as he put it, he wanted a formula for $\zeta(s)$ "which remains valid for all s." Such a formula would give the same values for $\zeta(s)$ as does Euler's definition (1.2.6) when $\sigma > 1$, but it would also give sensible values for $\zeta(s)$ even when $\sigma < 1$. Riemann was fabulously successful in discovering how to do that. He did it by discovering what is called the functional equation of the zeta function, and, just to anticipate things a bit, here it is (we'll have derived it by the end of this section):

$$\zeta(s) = 2(2\pi)^{s-1} \sin\left(\frac{\pi s}{2}\right) \Gamma(1-s)\zeta(1-s).$$

Riemann's functional equation of the zeta function is considered to be one of the gems of mathematics. Here's how it works. What we have is

$$\zeta(s) = F(s)\zeta(1-s), \quad F(s) = 2(2\pi)^{s-1} \sin\left(\frac{\pi s}{2}\right) \Gamma(1-s).$$

$F(s)$ is a well-defined function for all σ. So, if we have an s with $\sigma > 1$, we'll use Euler's formulation in (1.2.6) to compute $\zeta(s)$, but if $\sigma < 0$, we'll use the functional equation (along with Euler's formulation in (1.2.6) to compute the factor $\zeta(1 - s)$ on the right-hand side of the functional equation, because the real part of $1 - s$ is > 1 if $\sigma < 0$).

There is, of course, the remaining question of how to compute $\zeta(s)$ for the case of $0 < \sigma < 1$, for which s is in the so-called critical strip (a vertical band with width 1 extending from $-i\infty$ to $+i\infty$). The functional equation doesn't help us now, because if s is in the critical strip, then so is $1 - s$. This is actually a problem we've already solved, however, as you can see by looking back at the solution to Challenge Problem 2.1.3, which uses the eta function

$$\eta(s) = \sum_{k=1}^{\infty} \frac{(-1)^{k+1}}{k^s} = [1 - 2^{1-s}]\zeta(s).$$

So, for example, right in the middle of the critical strip, on the real axis, we have $s = 1/2$, and to repeat the Challenge Problem solution just a bit, we have

$$\sum_{k=1}^{\infty} \frac{(-1)^{k+1}}{k^{1/2}} = \left[1 - 2^{1-(\frac{1}{2})}\right]\zeta\left(\frac{1}{2}\right) = \left[\frac{1}{\sqrt{1}} - \frac{1}{\sqrt{2}} + \frac{1}{\sqrt{3}} - \frac{1}{\sqrt{4}} + \cdots\right].$$

Thus,

$$\zeta\left(\frac{1}{2}\right) = \frac{1}{1 - \sqrt{2}}\left[1 - \frac{1}{\sqrt{2}} + \frac{1}{\sqrt{3}} - \frac{1}{\sqrt{4}} + \cdots\right].$$

If we keep the first 1 million terms—a well-known theorem in freshman calculus tells us that any alternating series (with monotonically decreasing terms approaching zero) always converges (look back at note 6 in Chapter 1), and the maximum error made in using just a finite number of terms in a partial sum is less than the first term neglected—our error for a partial sum using 1 million terms should then be less than 10^{-3}, and we get

$$\zeta\left(\frac{1}{2}\right) \approx -1.46.$$

For Euler's case of s purely real, the plots (generated by *MATLAB*'s *zeta* function) in Figure 3.7.1 show the general behavior of $\zeta(s)$. (Compare the calculated value of $\zeta(\frac{1}{2})$ with the lower-right plot.) For $s > 1$, $\zeta(s)$ smoothly decreases from $+\infty$ toward 1 as s increases from 1, while for $s < 0$, $\zeta(s)$ oscillates, eventually heading off to $-\infty$ as s approaches 1 *from below*. Figure 3.7.1 indicates that $\zeta(0) = -0.5$ (a result we discussed in the previous chapter), and later in this section, I'll show you how to derive $\zeta(0) = -\frac{1}{2}$ using the functional equation. Figure 3.7.1 also shows that $\zeta'(0) < 0$, a result we already know from our exact calculation of $\zeta'(0)$ in (2.6.4). Notice, too, that Figure 3.7.1 hints at $\zeta(s) = 0$ for s a negative, even integer, another conclusion supported by the functional equation.

To make that last observation crystal clear, let's write $s = -2n$, where $n = 0, 1, 2, 3, \ldots$ Then the functional equation becomes

$$\zeta(-2n) = -2^{-2n} \pi^{-(2n+1)} \Gamma(1+2n)\sin(n\pi)\zeta(1+2n) = 0$$

because all of the factors on the right of the first equality are finite for all n, including $\sin(n\pi)$, which is, of course, zero for all integer n. We must exclude the case of $n = 0$, however, because then $\zeta(1 + 2n)$ $= \zeta(1) = \infty$, and this infinity is sufficient to overwhelm the zero value of $\sin(0)$. We know this because $\zeta(0) \neq 0$. When a value of s gives $\zeta(s) = 0$ then we call that value of s a zero of the zeta function. Thus, all the even, negative integers are zeros of $\zeta(s)$, and

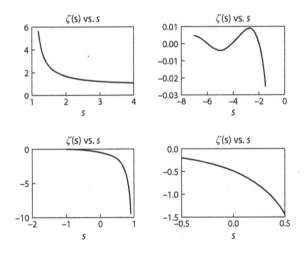

FIGURE 3.7.1.

The zeta function for real s.

because they are so easy to compute, they are called the trivial zeros of $\zeta(s)$. There are other zeros of $\zeta(s)$, however, which are not so easy to compute,[20] and where they are in the complex plane is what the RH is all about.

Here is what Riemann correctly believed about the non-trivial zeros (even if he couldn't prove all the following in 1859):

1. They are infinite in number.
2. All are complex (of the form $s = \sigma + it, t \neq 0$).
3. All are in the critical strip ($0 < \sigma < 1$).
4. They occur in pairs, symmetrically displaced around the vertical $\sigma = 1/2$ line (called the critical line), that is, if $\frac{1}{2} - \varepsilon + it$ is a zero for some t, then so is $\frac{1}{2} + \varepsilon + it$ for some ε in the interval $0 \leq \varepsilon < \frac{1}{2}$.
5. They are symmetrical about the real axis ($t = 0$); that is, if $\sigma + it$ is a zero, then so is $\sigma - it$ (the zeros appear as conjugate pairs).

The RH is now easy to state: $\varepsilon = 0$. That is, *all* of the complex zeros are on the critical line and so have a real part exactly equal to $\sigma = \frac{1}{2}$. As Riemann conjectured, "it is *very probable* [my emphasis] that all the [complex zeros are on the critical line]." Since 1859, all who have tried to prove the RH have failed, including Riemann, who wrote: "Certainly one would wish [for a proof]; I have meanwhile temporarily put aside the search for [a proof] after some fleeting futile attempts, as it appears unnecessary for [finding a formula for $\pi(x)$]" (see Edwards, *Riemann's Zeta Function*).

There does appear, at first glance, to be quite substantial computational support for the truth of the RH. Ever since Riemann himself

---

20. The methods used to compute the non-trivial zeros are far from obvious and are certainly beyond the level of this book. If you are interested in looking further into how such computations are done, I can recommend the following four books: (1) H. M. Edwards, *Riemann's Zeta Function* (Academic Press, 1974); (2) E. C. Titchmarsh, *The Theory of the Riemann Zeta-Function*, 2nd edition, revised by D. R. Heath-Brown (Oxford Science Publications, 1986); (3) Aleksandar Ivić, *The Riemann Zeta-Function* (John Wiley & Sons, 1985); and (4) Peter Borwein et al., editors, *The Riemann Hypothesis* (Springer, 2008).

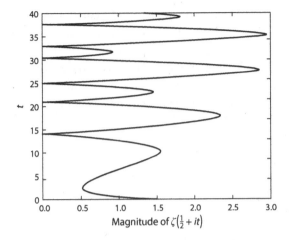

FIGURE 3.7.2.

The first six zeros of $\zeta(\frac{1}{2} + it)$.

hand computed the locations of the first three complex zeros,[21] with the aid of electronic computers, the past few decades have seen that accomplishment vastly surpassed. In 2004, the first 10^{13} (yes, 10 trillion!) zeros were shown to all be on the critical line. Since even a single zero off the critical line, by even the smallest amount, is all that is needed to disprove the RH, this looks pretty impressive—but mathematicians are, frankly, *not* impressed. As Ivić wrote in his book (see note 20), "No doubt the numerical data will continue to accrue, but number theory is unfortunately one of the branches of mathematics where numerical evidence does not count for much."

21. Because of the symmetry properties of the complex zero locations, one only has to consider the case of $t > 0$. The value of t for a zero is called the *height* of the zero, and the zeros are ordered by increasing height. The first six zeros are shown in Figure 3.7.2, where a zero occurs each place $|\zeta(\frac{1}{2}+it)|$ touches the vertical t-axis. (The horizontal axis is the magnitude of $\zeta(\frac{1}{2}+it)$.) The heights of the first six zeros are 14.134725, 21.022040, 25.010856, 30.424878, 32.935057, and 37.586176. In addition to the first 10^{13} zeros, billions more zeros at heights as large as 10^{24} have also been confirmed to all be on the critical line.

There are, in fact, lots of historical examples in mathematics where initial, seemingly massive computational "evidence" has prompted conjectures that later proved to be false. A particularly famous example involves $\pi(x)$ and $\text{li}(x)$. For all values of x for which $\pi(x)$ and $\text{li}(x)$ are known, $\pi(x) < \text{li}(x)$. Furthermore, the difference between the two increases as x increases and, for large x, the difference is significant; for $x = 10^{18}$, for example, $d(x) = \text{li}(x) - \pi(x) \approx$ 22,000,000. Based on this impressive numerical "evidence," it was commonly believed, for a long time, that $d(x) > 0$ for *all* x. Gauss believed this (as did Riemann) all his life. *But it's not true.*

In 1912, Hardy's friend and collaborator J. E. Littlewood (1885–1977) proved that there is some x for which $d(x) < 0$. Two years later, he extended his proof to show that as x continues to increase, the sign of $d(x)$ flips back and forth endlessly. The value of x at which the first change in sign of $d(x)$ occurs is not known, only that it is very big. In 1933 Littlewood's student, the South African Stanley Skewes (1899–1988), derived a stupendously huge upper bound on the value of that first $x : e^{e^{e^{79}}} \approx 10^{10^{10^{34}}}$. This has become famous in mathematics as the first Skewes number. In his derivation, Skewes assumed the truth of the RH, but in 1955 he dropped that assumption to calculate a new upper bound for the first x at which $d(x)$ changes sign: This is the second Skewes number, equal to $10^{10^{10^{1,000}}}$, and it is *tremendously* larger than the first one. (In 2000 the upper bound was reduced to "just" 1.39×10^{316}.) All of these numbers are far beyond anything that can be numerically studied on a computer, and the trillions of complex zeros that have all been found on the critical line are minuscule in number in comparison. It is entirely possible that the first complex zero off the critical line (thus disproving the RH) may not occur until a vastly greater height is reached than has been examined so far.

Some mathematicians have been markedly less than enthusiastic about the RH. Littlewood, in particular, was quite blunt, writing "I believe [the RH] to be false. There is no evidence for it. . . . One should not believe things for which there is no evidence. . . . I have discussed the matter with several people who know the problem in relation to electronic calculation; they are all agreed that the chance of finding a zero off the line in a lifetime's calculation is millions to one against it. It looks then as if we may never know."[22] A slightly more muted (but perhaps not by much) position is that of the American mathematician H. M. Edwards (1936–2020), who wrote in his classic book on the zeta function (see his *Riemann's Zeta Function*): "Riemann based his hypothesis on no insights . . . which are not available to us today . . . and that, on the contrary, had he known some of the facts which have since been discovered, he might well have been led to reconsider . . . unless some basic cause is operating which has eluded mathematicians for 110 years [160 years now, as I write in 2020], occasional [complex zeros] off the [critical] line are altogether possible. . . . Riemann's insight was stupendous, but it was not supernatural, and what seemed 'probable' to him in 1859 might seem less so today."

Okay, that's a short history of the RH, but more to the point here, where does the functional equation of the zeta function come from? The derivation of the functional equation for $\zeta(s)$ that appears in Riemann's famous 1859 paper uses the advanced math of complex function theory, which is just beyond the level of this book. So, what I'll now show you is a different derivation (but one also due to Riemann) that makes clever use of nothing but AP-calculus. We start with the integral

$$\int_0^\infty x^{m-1} e^{-ax} dx, m \geq 1, a > 0,$$

and make the change of variable $u = ax$ (and so $dx = du/a$). Thus,

$$\int_0^\infty x^{m-1} e^{-ax}\, dx = \int_0^\infty \left(\frac{u}{a}\right)^{m-1} e^{-u} \frac{du}{a} = \frac{1}{a^m} \int_0^\infty u^{m-1} e^{-u}\, du.$$

22. From Littlewood's essay, "The Riemann Hypothesis," in I. J. Good, editor, *The Scientist Speculates: An Anthology of Partly-Baked Ideas* (Basic Books, 1962).

The right-most integral is, from (1.4.1), $\Gamma(m)$, and so

(3.7.1)
$$\int_0^\infty x^{m-1} e^{-ax} dx = \frac{\Gamma(m)}{a^m}.$$

Now, if we let

$$m - 1 = \frac{1}{2}s - 1 \left(\text{that is, } m = \frac{1}{2}s \right)$$

and

$$a = n^2 \pi,$$

then (3.7.1) becomes

(3.7.2)
$$\int_0^\infty x^{\frac{1}{2}s-1} e^{-n^2\pi x} dx = \frac{\Gamma\left(\frac{1}{2}s\right)}{(n^2\pi)^{\frac{1}{2}s}} = \frac{\Gamma\left(\frac{1}{2}s\right)}{\pi^{\frac{1}{2}s} n^s}.$$

Then, summing (3.7.2) over all positive integer n, we have

$$\sum_{n=1}^\infty \int_0^\infty x^{\frac{1}{2}s-1} e^{-n^2\pi x} dx = \sum_{n=1}^\infty \frac{\Gamma\left(\frac{1}{2}s\right)}{\pi^{\frac{1}{2}s} n^s}$$

or, reversing the order of summation and integration on the left,

(3.7.3)
$$\int_0^\infty x^{\frac{1}{2}s-1} \sum_{n=1}^\infty e^{-n^2\pi x} dx$$

$$= \pi^{-\frac{1}{2}s} \Gamma\left(\frac{1}{2}s\right) \sum_{n=1}^\infty \frac{1}{n^s} = \pi^{-\frac{1}{2}s} \Gamma\left(\frac{1}{2}s\right) \zeta(s).$$

At this point Riemann defined the function

$$(3.7.4) \qquad \psi(x) = \sum_{n=1}^{\infty} e^{-n^2 \pi x},$$

and then used the identity[23]

$$(3.7.5) \qquad \sum_{n=-\infty}^{\infty} e^{-n^2 \pi x} = \frac{1}{\sqrt{x}} \sum_{n=-\infty}^{\infty} e^{-n^2 \pi / x}.$$

The left-hand side of (3.7.5) is (because $n^2 > 0$ for n negative *or* positive) $\sum_{n=-\infty}^{-1} e^{-n^2 \pi x} + 1 + \sum_{n=1}^{\infty} e^{-n^2 \pi x} = \sum_{n=1}^{\infty} e^{-n^2 \pi x} + 1 + \sum_{n=1}^{\infty} e^{-n^2 \pi x} = 2\psi(x) + 1.$ The right-hand side of (3.7.5) is (for the same reason)

$$\frac{1}{\sqrt{x}} \sum_{n=-\infty}^{\infty} e^{-n^2 \pi / x} = \frac{1}{\sqrt{x}} \left\{ 2\psi\left(\frac{1}{x}\right) + 1 \right\}.$$

Thus,

$$2\psi(x) + 1 = \frac{1}{\sqrt{x}} \left\{ 2\psi\left(\frac{1}{x}\right) + 1 \right\}$$

or, solving for $\psi(x)$,

$$(3.7.6) \qquad \psi(x) = \frac{1}{\sqrt{x}} \psi\left(\frac{1}{x}\right) + \frac{1}{2\sqrt{x}} - \frac{1}{2} = \sum_{n=1}^{\infty} e^{-n^2 \pi x}.$$

23. Recall that we derived the identity $\sum_{k=-\infty}^{\infty} e^{-ak^2} = \sqrt{\frac{\pi}{a}} \sum_{n=-\infty}^{\infty} e^{-\pi^2 n^2 / a}$ in (3.6.18). If you write $a = \pi x$, then (3.7.5) immediately results.

Now, putting (3.7.4) into (3.7.3) gives us

$$\pi^{-\frac{1}{2}s}\Gamma\left(\frac{1}{2}s\right)\zeta(s)=\int_0^\infty x^{\frac{1}{2}s-1}\psi(x)dx$$

or, breaking the integral into two parts,

$$(3.7.7)\quad \pi^{-\frac{1}{2}s}\Gamma\left(\frac{1}{2}s\right)\zeta(s)=\int_0^1 x^{\frac{1}{2}s-1}\psi(x)dx+\int_1^\infty x^{\frac{1}{2}s-1}\psi(x)dx.$$

Using (3.7.6) in the first integral on the right of (3.7.7), we have

$$\pi^{-\frac{1}{2}s}\Gamma\left(\frac{1}{2}s\right)\zeta(s)=\int_0^1 x^{\frac{1}{2}s-1}\left\{\frac{1}{\sqrt{x}}\psi\left(\frac{1}{x}\right)+\frac{1}{2\sqrt{x}}-\frac{1}{2}\right\}dx$$

$$+\int_1^\infty x^{\frac{1}{2}s-1}\psi(x)dx$$

$$=\int_0^1 x^{\frac{1}{2}s-1}\left\{\frac{1}{2\sqrt{x}}-\frac{1}{2}\right\}dx+\int_0^1 x^{\frac{1}{2}s-\frac{3}{2}}\psi\left(\frac{1}{x}\right)dx+\int_0^\infty x^{\frac{1}{2}s-1}\psi(x)dx.$$

The first integral on the right is easy to do (for $s>1$):

$$\int_0^1 x^{\frac{1}{2}s-1}\left\{\frac{1}{2\sqrt{x}}-\frac{1}{2}\right\}dx=\frac{1}{2}\int_0^1 x^{\frac{1}{2}s-\frac{3}{2}}dx-\frac{1}{2}\int_0^1 x^{\frac{1}{2}s-1}dx$$

$$=\frac{1}{2}\left(\frac{x^{\frac{1}{2}s-\frac{1}{2}}}{\frac{1}{2}s-\frac{1}{2}}\right)\bigg|_0^1-\frac{1}{2}\left(\frac{x^{\frac{1}{2}s}}{\frac{1}{2}s}\right)\bigg|_0^1=\frac{1}{2}\left(\frac{1}{\frac{1}{2}s-\frac{1}{2}}\right)$$

$$-\frac{1}{2}\left(\frac{1}{\frac{1}{2}s}\right)=\frac{1}{s-1}-\frac{1}{s}=\frac{1}{s(s-1)}.$$

Thus,

$$(3.7.8) \quad \pi^{-\frac{1}{2}s}\Gamma\left(\frac{1}{2}s\right)\zeta(s) = \frac{1}{s(s-1)} + \int_0^1 x^{\frac{1}{2}s-\frac{3}{2}}\psi\left(\frac{1}{x}\right)dx + \int_1^\infty x^{\frac{1}{2}s-1}\psi(x)\,dx.$$

Next, in the first integral on the right in (3.7.8), make the change of variable $u = \frac{1}{x}$ (and so $dx = -\frac{du}{u^2}$). Then

$$\int_0^1 x^{\frac{1}{2}s-\frac{3}{2}}\psi(\tfrac{1}{x})dx = \int_\infty^1 (\tfrac{1}{u})^{\frac{1}{2}s-\frac{3}{2}}\psi(u)\{-\tfrac{du}{u^2}\} = \int_1^\infty \frac{1}{u^{\frac{1}{2}s+\frac{1}{2}}}\psi(u)du = \int_1^\infty x^{-\frac{1}{2}s-\frac{1}{2}}\psi(x)\,dx$$

and therefore (3.7.8) becomes

$$(3.7.9) \quad \pi^{-\frac{1}{2}s}\Gamma\left(\frac{1}{2}s\right)\zeta(s) = \frac{1}{s(s-1)} + \int_1^\infty \left\{ x^{-\frac{1}{2}s-\frac{1}{2}} + x^{\frac{1}{2}s-1} \right\}\psi(x)dx.$$

All we need do now is notice, as did Riemann, that the right-hand side of (3.7.9) is unchanged if we replace every occurrence of s with $1 - s$. Try it and see. But that means we can do the same thing on the left-hand side of (3.7.9), because, after all, (3.7.9) is an identity. That is, it must be true that

$$(3.7.10) \quad \pi^{-\frac{1}{2}s}\Gamma\left(\frac{1}{2}s\right)\zeta(s) = \pi^{-\frac{1}{2}(1-s)}\Gamma\left(\frac{1-s}{2}\right)\zeta(1-s).$$

We are now almost done, with just a few more routine steps to go.

Solving (3.7.10) for $\zeta(s)$, we have

$$(3.7.11) \quad \zeta(s) = \pi^{s-\frac{1}{2}}\frac{\Gamma\left(\dfrac{1-s}{2}\right)}{\Gamma\left(\dfrac{1}{2}s\right)}\zeta(1-s).$$

Now, recall (2.2.7), one of the forms of Legendre's duplication formula:

$$z!\left(z-\frac{1}{2}\right)! = 2^{-2z}\pi^{\frac{1}{2}}(2z)!$$

or, expressed in gamma notation,

(3.7.12) $$\Gamma(z+1)\Gamma\left(z+\frac{1}{2}\right) = 2^{-2z}\pi^{\frac{1}{2}}\Gamma(2z+1).$$

If we write $2z + 1 = 1 - s$, then $z = -s/2$, and (3.7.12) becomes

$$\Gamma\left(-\frac{s}{2}+1\right)\Gamma\left(-\frac{s}{2}+\frac{1}{2}\right) = 2^{s}\pi^{\frac{1}{2}}\Gamma(1-s) = \Gamma\left(1-\frac{s}{2}\right)\Gamma\left(\frac{1-s}{2}\right)$$

or

(3.7.13) $$\Gamma\left(\frac{1-s}{2}\right) = \frac{2^{s}\pi^{\frac{1}{2}}\Gamma(1-s)}{\Gamma\left(1-\frac{s}{2}\right)}.$$

From (1.4.20), the reflection formula for the gamma function,

$$\Gamma(m)\Gamma(1-m) = \frac{\pi}{\sin(m\pi)}$$

or, with $m = \frac{s}{2}$,

$$\Gamma\left(\frac{s}{2}\right)\Gamma\left(1-\frac{s}{2}\right) = \frac{\pi}{\sin\left(\dfrac{\pi s}{2}\right)}$$

and this says that

(3.7.14)
$$\Gamma\left(1-\frac{s}{2}\right) = \frac{\pi}{\Gamma\left(\dfrac{s}{2}\right)\sin\left(\dfrac{\pi s}{2}\right)}.$$

So, putting (3.7.14) into (3.7.13), we have

$$\Gamma\left(\frac{1-s}{2}\right) = \frac{2^{s}\pi^{\frac{1}{2}}\Gamma(1-s)}{\Gamma\left(\dfrac{s}{2}\right)\sin\left(\dfrac{\pi s}{2}\right)} = 2^{s}\pi^{-\frac{1}{2}}\Gamma\left(\frac{s}{2}\right)\sin\left(\frac{\pi s}{2}\right)\Gamma(1-s)$$

or

(3.7.15)
$$\frac{\Gamma\left(\dfrac{1-s}{2}\right)}{\Gamma\left(\dfrac{s}{2}\right)} = 2^{s}\pi^{-\frac{1}{2}}\sin\left(\frac{\pi s}{2}\right)\Gamma(1-s).$$

Inserting (3.7.15) into (3.7.11), we arrive at

$$\zeta(s) = \pi^{s-\frac{1}{2}}2^{s}\pi^{-\frac{1}{2}}\sin\left(\frac{\pi s}{2}\right)\Gamma(1-s)\zeta(1-s)$$

and so, at last, we finally have our prize:

(3.7.16)
$$\zeta(s) = 2(2\pi)^{s-1}\sin\left(\frac{\pi s}{2}\right)\Gamma(1-s)\zeta(1-s),$$

which is the functional equation of the zeta function.

As a simple test of (3.7.16), suppose $s = \frac{1}{2}$. Then

$$\zeta\left(\frac{1}{2}\right) = 2(2\pi)^{-\frac{1}{2}} \sin\left(\frac{\pi}{4}\right) \Gamma\left(\frac{1}{2}\right) \zeta\left(\frac{1}{2}\right),$$

which says, once we cancel the $\zeta(\frac{1}{2})$ on each side,[24] that

$$1 = \frac{2}{\sqrt{2\pi}} \sin\left(\frac{\pi}{4}\right) \Gamma\left(\frac{1}{2}\right).$$

Is this correct? Yes, because the right-hand side is

$$\left(\sqrt{\frac{2}{\pi}}\right)\left(\frac{1}{\sqrt{2}}\right)\left(\sqrt{\pi}\right) = 1.$$

So, (3.7.16) is consistent for $s = \frac{1}{2}$.

As an example of how (3.7.16) works, let's use it to calculate $\zeta(-1)$. Thus, with $s = -1$,

$$\zeta(-1) = 2(2\pi)^{-2} \sin\left(-\frac{\pi}{2}\right) \Gamma(2) \zeta(2).$$

Since $\Gamma(2) = 1$, $\sin\left(-\frac{\pi}{2}\right) = -1$, and $\zeta(2) = \frac{\pi^2}{6}$, then

$$\zeta(-1) = \frac{2}{4\pi^2}(-1)\frac{\pi^2}{6} = -\frac{1}{12},$$

just as Ramanujan got (with far more by-guess-by-gosh-anything-goes symbol pushing) in (2.1.11).

As a second illustration of (3.7.16) in action, I'll use it next to calculate the value of $\zeta(0)$, which of course we already know from

---

24. We know we can do this, because as we determined earlier by direct computation, $\zeta\left(\frac{1}{2}\right) \approx -1.46\cdots \neq 0$.

Challenge Problem 2.1.3. If we do something as crude as just shove $s = 0$ into (3.7.16), we quickly see that we get nowhere:

$$\zeta(0) = 2(2\pi)^{-1}\sin(0)\Gamma(1)\zeta(1) = ?,$$

because the zero of $\sin(0)$ and the infinity of $\zeta(1)$ are at war with each other. Which one wins? To find out, we'll have to be a lot more subtle in our calculations. Strange as it may at first seem, we'll get our answer by studying the case of $s = 1$ (not $s = 0$), which I'll simply ask you to take on faith as we start.

Looking back at (3.7.9), we have

$$\zeta(s) = \frac{1}{\pi^{-\frac{1}{2}s}\Gamma\left(\frac{1}{2}s\right)s(s-1)} + \frac{1}{\pi^{-\frac{1}{2}s}\Gamma\left(\frac{1}{2}s\right)}\int_1^{\infty}\left\{x^{-\frac{1}{2}s-\frac{1}{2}} + x^{\frac{1}{2}s-1}\right\}\psi(x)dx.$$

If we let $s \to 1$, then we see that the right-hand side does indeed blow up (as it should, because $\zeta(1) = \infty$), strictly because of the first term on the right, alone, since the integral term is obviously convergent.[25] In fact, since $\lim_{s \to 1}\pi^{-\frac{1}{2}s}\Gamma(\frac{1}{2}s)s = \frac{\Gamma(\frac{1}{2})}{\sqrt{\pi}} = \frac{\sqrt{\pi}}{\sqrt{\pi}} = 1$, then $\zeta(s)$ blows up like $1/(s-1)$ as $s \to 1$. *Remember this point—it will prove to be the key to our solution.* (You'll recall we already know this, from our work in Section 2.5.)

Now, from (3.7.16) we have

$$\zeta(1-s) = \frac{\zeta(s)}{2(2\pi)^{s-1}\sin\left(\dfrac{\pi s}{2}\right)\Gamma(1-s)}.$$

---

25. I use the word "obviously" because, over the entire interval of integration, the integrand is finite and goes to zero very fast as $x \to \infty$. Indeed, the integrand vanishes even faster than exponentially as $x \to \infty$, which you can show by using (3.7.4) to write $\psi(x) = \sum_{n=1}^{\infty}e^{-n^2\pi x} = e^{-\pi x} + e^{-4\pi x} + e^{-9\pi x} + \cdots < e^{-\pi x} + e^{-2\pi x} + e^{-3\pi x} + \cdots$, a geometric series easily summed to give $\psi(x) < \frac{1}{e^{\pi x}-1}, x > 0$, which behaves like $e^{-\pi x}$ for x large. With $s = 1$ the integrand behaves (for x large) like $\frac{\psi(x)}{x^{3/2}+x^{1/2}} \approx \frac{e^{-\pi x}}{x\sqrt{x}}$ for x large.

From the reflection formula for the gamma function (1.4.20), we have

$$\Gamma(s)\Gamma(1-s)=\frac{\pi}{\sin(\pi s)}$$

and so

$$\Gamma(1-s)=\frac{\pi}{\Gamma(s)\sin(\pi s)},$$

which says

$$\zeta(1-s)=\frac{\zeta(s)}{2(2\pi)^{s-1}\sin\left(\dfrac{\pi s}{2}\right)}\frac{\pi}{\Gamma(s)\sin(\pi s)}=\frac{\Gamma(s)\sin(\pi s)\zeta(s)}{2\pi(2\pi)^{s-1}\sin\left(\dfrac{\pi s}{2}\right)}.$$

Since $\sin(\pi s)=2\cos(\frac{\pi s}{2})\sin(\frac{\pi s}{2})$, we arrive at

$$(3.7.17)\qquad \zeta(1-s)=\frac{\Gamma(s)\cos\left(\dfrac{\pi s}{2}\right)\zeta(s)}{\pi(2\pi)^{s-1}},$$

an alternative form of the functional equation for the zeta function. This is the form we'll use to let $s \to 1$, thus giving $\zeta(0)$ on the left.

So, from (3.7.17) we have

$$\lim_{s\to 1}\zeta(1-s)=\zeta(0)=\lim_{s\to 1}\frac{\Gamma(s)\cos\left(\dfrac{\pi s}{2}\right)\zeta(s)}{\pi(2\pi)^{s-1}}=\frac{\Gamma(1)}{\pi}\lim_{s\to 1}\cos\left(\dfrac{\pi s}{2}\right)\zeta(s)$$

$$= \frac{1}{\pi} \lim_{s \to 1} \cos\left(\frac{\pi s}{2}\right) \zeta(s)$$

or

(3.7.18)
$$\zeta(0) = \frac{1}{\pi} \lim_{s \to 1} \frac{\cos\left(\dfrac{\pi s}{2}\right)}{s-1}$$

where I've used our earlier conclusion that $\zeta(s)$ behaves like $\frac{1}{s-1}$ as $s \to 1$. The limit in (3.7.18) gives the indeterminate result $\frac{0}{0}$, and so we use L'Hôpital's rule to compute

$$\zeta(0) = \frac{1}{\pi} \lim_{s \to 1} \frac{\dfrac{d}{ds}\left\{\cos\left(\dfrac{\pi s}{2}\right)\right\}}{\dfrac{d}{ds}\{s-1\}} = \frac{1}{\pi} \lim_{s \to 1} \frac{-\dfrac{\pi}{2} \sin\left(\dfrac{\pi s}{2}\right)}{1}$$

and, at last, we have our answer:

$$\zeta(0) = -\frac{1}{2}.$$

Now, here's a fitting note on which to end this chapter on amazing mathematics: In 1749, more than a century before Riemann, Euler *guessed* (3.7.17)! As he wrote,[26] "I shall hazard the following conjecture [and then follows the equivalent of (3.7.17)]." Riemann was the first to prove the functional equation, yes, but Euler was the first to know it.

26. See Raymond Ayoub, "Euler and the Zeta Function," *American Mathematical Monthly*, December 1974, pp. 1067–1086.

Challenge Problem 3.7.1: If $s = x + it$, $x > 1$, find expressions (as infinite sums) for the real and imaginary parts of $\zeta(s)$. Use your expressions to calculate the value of $\zeta(2 + i)$. Hint: A look back at Challenge Problem 1.1.1 may be helpful.

Challenge Problem 3.7.2: Calculate the numerical value of $\zeta(-\frac{1}{2})$ to at least four decimal places. Compare your answer to the lower-right plot of Figure 3.7.1. Hint: Start with $s = -\frac{1}{2}$ and then use the functional equation of the zeta function.

Euler Sums, the Harmonic Series, and the Zeta Function

4.1 Euler's Original Sums

From Chapter 1 you'll recall (1.2.3):

$$h(q) = \sum_{k=1}^{q} \frac{1}{k} = 1 + \frac{1}{2} + \frac{1}{3} + \cdots + \frac{1}{q}, \ q \geq 1.$$

In 1775 Euler arrived at the following two amazing expressions (now called Euler sums) involving $h(q)$ and values of the zeta function:

(4.1.1)
$$\sum_{q=1}^{\infty} \frac{h(q)}{q^2} = 2\zeta(3) = 2.404113\ldots$$

and

(4.1.2) $$\sum_{q=1}^{\infty} \frac{h(q)}{q^3} = \frac{5}{4}\zeta(4) = \frac{\pi^4}{72} = 1.352904\ldots.$$

Are you surprised to see π^4 come out of the sum in (4.1.2)? Was Euler right? Well, he was, after all, *Euler*, and so it's a lot better than an even bet he was right, but even Euler liked to check his work with numerical confirmation. Still, even for Euler, doing the sums in (4.1.1) and (4.1.2) for even a small number of terms must have been pretty grubby work, but today we can easily and quickly convince ourselves that (4.1.1) and (4.1.2) are "pretty likely" correct. I am speaking, of course, of directly computing those sums on an electronic computer.

Indeed, those are simple enough calculations, in principle, that they could be programmed on a hand-held calculator. The following box shows how to code (4.1.1) using just a couple of loops (the code **euler1775.m** is in *MATLAB*, but equivalent codes in other scientific programming languages would look quite similar). This is an engineering approach that many mathematicians view with various degrees of regret, but let's do it anyway.

Figure 4.1.1 shows a semi-log plot of the partial sums of (4.1.1)—created by the last four lines of **euler1775.m**—using the first 10 million terms of (4.1.1). The 10-millionth partial sum is 2.404112, and so it certainly seems that (4.1.1) is plausible. A similar numerical evaluation of the sum of (4.1.2), shown in Figure 4.1.2, gives the value of the partial sum of the first 10 million terms as 1.352904 (in **euler1775.m** simply change the line ending in /q^2; to /q^3;). This is in excellent agreement with the right-hand side of (4.1.2).

```
%euler1775.m
q=1;h(q)=1;
for q=2:10000000
    h(q)=h(q-1)+1/q;
end
```

```
s(1)=1;
for q=2:10000000
    s(q)=s(q-1)+h(q)/q^2;
end
x=[1:1:10000000];
semilogx(x,s,'-k')
xlabel('number of terms')
ylabel('sum')
```

The sums of (4.1.1) and (4.1.2) are the $n = 2$ and $n = 3$ special cases, respectively, of the following recursive formula discovered by Euler in 1775:

(4.1.3) $\quad 2\sum_{q=1}^{\infty}\frac{h(q)}{q^n}=(n+2)\zeta(n+1)-\sum_{k=1}^{n-2}\zeta(n-k)\zeta(k+1),\ n\ge 2.$

The very next case ($n = 4$) of (4.1.3) gives (you should verify this)

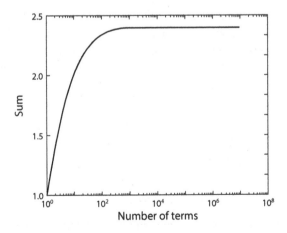

FIGURE 4.1.1.

The convergence of (4.1.1).

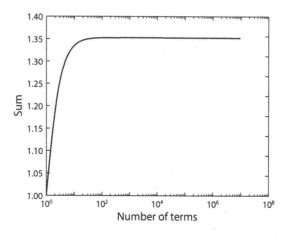

The convergence of (4.1.2).

(4.1.4) $$\sum_{q=1}^{\infty} \frac{h(q)}{q^4} = 3\zeta(5) - \zeta(3)\zeta(2).$$

We can numerically check (4.1.4) with the code **euler1775.m** by simply changing the line ending in /q^2; to /q^4;. The partial sum, using the first 10 million terms of (4.1.4), is 1.1334789 . . . , while if we type *3*zeta(5)-zeta(3)*zeta(2)* into *MATLAB,* we get the value 1.1334789 . . . . I think it's fair to say this is pretty good agreement.

Challenge Problem 4.1.1: If you plug $n = 3$ into Euler's recursion formula (4.1.3), you get $\sum_{q=1}^{\infty} \frac{h(q)}{q^3} = \frac{5}{2}\zeta(4) - \frac{1}{2}\zeta^2(2)$, an expression that doesn't look like the right-hand side of (4.1.2). Despite that, show the two expressions are, in fact, equal.

4.2 The Algebra of Euler Sums

The computer calculations of the previous section greatly increase our enthusiasm for Euler's (4.1.1) and (4.1.2), but how does one derive them in the first place? They are simply too bizarre to have just been made up. Amazingly, both of Euler's sums can be derived using math no more advanced than first-year high school algebra (no calculus, with the exception of one step that I think you'll accept as plausible—and then later in the book I'll show you how to establish that step with freshman calculus). It's all very clever algebra, to be sure, but nothing a bright teenager can't follow. We start by writing $f(n)$ as denoting some sequence of values that, as $n \to \infty$, approaches a finite limit $f(\infty) = L$. For example, if $f(n) = 1/n$, then $L = 0$.

Next, for integers M and q, where $M \geq q \geq 1$, we form the sum

$$(4.2.1) \quad \sum_{n=1}^{M} \{ f(n) - f(n+q) \} = [\{ f(1) + f(2) + f(3) + \cdots + f(M) \}$$

$$- \{ f(1+q) + f(2+q) + f(3+q) + \cdots + f(M+q) \}].$$

The first q terms in the first pair of curly brackets on the right of (4.2.1) survive the subtraction operation, but all the rest of the terms from $f(1 + q)$ to $f(M)$ are canceled by the same terms in the second pair of curly brackets. All of the remaining terms in the second pair of curly brackets, from $f(M + 1)$ to $f(M + q)$, remain. So, (4.2.1) becomes

$$(4.2.2) \quad \sum_{n=1}^{M} \{ f(n) - f(n+q) \} = \sum_{n=1}^{q} f(n) - \sum_{n=1}^{q} f(M+n).$$

The algebraic discussion you're reading here is due to the Indian Ankur Basu and the American Tom Apostol (whom you'll recall from Sections 2.4 and 2.6), as published in their paper, "A New Method for Investigating Euler Sums," *The Ramanujan Journal*, December 2000, pp. 397–419. This paper has a provocative footnote that appears at the bottom of the first page: "Translated from a handwritten manuscript, revised, edited, and prepared for

publication by the second author." The "second author" was
Apostol, a famous math professor at Caltech who died in 2016.
The first-named author, however, was an unknown, with his listed
address being simply that of the private residence of a relative's
home in West Bengal, India. As I read that footnote, I instantly
imagined a well-traveled envelope (plastered, I further imagined,
with exotic mailing stamps) unexpectedly appearing one morning
in Apostol's mailbox in the math department at Caltech. Was this,
as my perhaps now overheated imagination roared full-speed
ahead, a reenactment of Ramanujan's discovery, nearly a century
earlier, by G. H. Hardy (look back at Section 1.5)? It's too late to
ask Apostol about that, or about his emotional state as he tore
open that envelope and began to read its handwritten contents,
but Basu (as I write in 2020) appears on the Web as affiliated with
the faculty of the Department of Industrial Economics and
Management at the KTH Royal Institute of Technology in
Stockholm, Sweden. I think there just has to be a romantic story
about the origin of the Basu/Apostol paper, and perhaps Basu will
one day tell that story.

If we now let $M \to \infty$ we see that the first sum on the right of (4.2.2)
is unaffected, while all the terms in the second sum are equal to
$f(\infty) = L$. That is, for $f(n) = 1/n$ where $L = 0$, we have (after letting
$M \to \infty$)

$$(4.2.3) \qquad \sum_{n=1}^{\infty} \{f(n) - f(n+q)\} = \sum_{n=1}^{q} f(n), \ q \geq 1.$$

Now, temporarily put (4.2.3) aside (but not for long, as we'll
soon return to it) and turn your attention to the sum

$$(4.2.4) \qquad \sum_{\substack{n=1 \\ n \neq q}}^{\infty} \{f(n-q) - f(n)\}, q \geq 1.$$

To study this sum, we'll separate (4.2.4) into two cases: (1) where we'll consider the terms with $n < q$, and (2) where we'll consider the terms with $n > q$. (There are, of course, no terms such that $n = q$ because of the $n \neq q$ condition.) If $f(n)$ is an *odd* sequence of values, which means $-f(-n) = f(n)$, as is, for example, $f(n) = 1/n$, then for case (1) we have $n - q < 0$ for $n = 1$ to $q - 1$, and so

$$\sum_{n=1}^{q-1}\{f(n-q)-f(n)\} = \sum_{n=1}^{q-1}\{-f(q-n)-f(n)\} = -\sum_{n=1}^{q-1}\{f(q-n)+f(n)\}.$$

Since $q - n$ runs through the same values as does n, as n goes from 1 to $q - 1$, we have, for case (1),

$$\sum_{n=1}^{q-1}\{f(n-q)-f(n)\} = -2\sum_{n=1}^{q-1}f(n)$$

or, if we run the summation index for the right-hand side sum up one additional value from $q - 1$ to q,

$$(4.2.5) \qquad \sum_{n=1}^{q-1}\{f(n-q)-f(n)\} = -2\sum_{n=1}^{q}f(n)+2f(q).$$

This takes care of case (1), $n < q$.

For case (2), we have n going from $q + 1$ to infinity. That is, we are now looking at

$$\sum_{n=q+1}^{\infty}\{f(n-q)-f(n)\}.$$

Let's define a new summation index $k = n - q$, and so k goes from 1 to infinity, and our sum becomes, for case (2),

$$\sum_{k=1}^{\infty}\{f(k)-f(k+q)\}.$$

Since the symbol we use for the index is arbitrary, let's go back to n and write our sum as

$$\sum_{n=1}^{\infty}\{f(n)-f(n+q)\}.$$

But we've already evaluated this sum, back in (4.2.3). So, for case (2) we can write

(4.2.6) $$\sum_{n=q+1}^{\infty}\{f(n-q)-f(n)\}=\sum_{n=1}^{q}f(n).$$

Our results in (4.2.5) and (4.2.6) can be combined to give us the value of the sum in (4.2.4): For an odd $f(n)$ such that $\lim_{n\to\infty}f(n)=0$ we have

$$\sum_{\substack{n=1\\n\neq q}}^{\infty}\{f(n-q)-f(n)\}=\left[-2\sum_{n=1}^{q}f(n)+2f(q)\right]+\left[\sum_{n=1}^{q}f(n)\right]$$

or

(4.2.7) $$\sum_{\substack{n=1\\n\neq q}}^{\infty}\{f(n-q)-f(n)\}=2f(q)-\sum_{n=1}^{q}f(n).$$

Now we are ready to derive Euler's two sums.

Explicitly writing $f(n)=1/n$ in (4.2.7), we have

$$\sum_{\substack{n=1\\n\neq q}}^{\infty}\left\{\frac{1}{n-q}-\frac{1}{n}\right\}=2\frac{1}{q}-\sum_{n=1}^{q}\frac{1}{n}.$$

Since

$$\frac{1}{n-q}-\frac{1}{n}=\frac{q}{n(n-q)}$$

and since (by definition)

$$\sum\nolimits_{n=1}^{q} \frac{1}{n} = h(q),$$

we have

$$\sum\nolimits_{\substack{n=1 \\ n \neq q}}^{\infty} \frac{q}{n(n-q)} = \frac{2}{q} - h(q)$$

or, dividing through by q, this becomes

(4.2.8) $$\sum\nolimits_{\substack{n=1 \\ n \neq q}}^{\infty} \frac{1}{n(n-q)} = \frac{2}{q^2} - \frac{h(q)}{q}.$$

Next, divide through (4.2.8) by q again and then sum over all q to get

$$\sum\nolimits_{q=1}^{\infty} \sum\nolimits_{\substack{n=1 \\ n \neq q}}^{\infty} \frac{1}{qn(n-q)} = 2 \sum\nolimits_{q=1}^{\infty} \frac{1}{q^3} - \sum\nolimits_{q=1}^{\infty} \frac{h(q)}{q^2}.$$

The first sum on the right is $\zeta(3)$, and so

(4.2.9) $$\sum\nolimits_{q=1}^{\infty} \frac{h(q)}{q^2} = 2\zeta(3) - \sum\nolimits_{q=1}^{\infty} \sum\nolimits_{\substack{n=1 \\ n \neq q}}^{\infty} \frac{1}{qn(n-q)}.$$

Euler's sum in (4.1.1) *is* (4.2.9) if we can argue that the double sum is zero. Is it? Yes. *Why?* Think about this for a while, and then I'll show you (in the box at the end of this section) a simple plausibility argument for why the double sum does indeed vanish.

Now, what about (4.1.2), Euler's other sum? Looking back at (4.2.3), if we set $f(n) = 1/n$ we get

$$\sum\nolimits_{n=1}^{\infty} \left\{ \frac{1}{n} - \frac{1}{n+q} \right\} = \sum\nolimits_{n=1}^{q} \frac{1}{n} = h(q) = \sum\nolimits_{n=1}^{\infty} \frac{q}{n(n+q)}$$

and so, dividing through by q,

(4.2.10) $$\sum_{n=1}^{\infty} \frac{1}{n(n+q)} = \frac{h(q)}{q}.$$

If we now impose the $n \neq q$ condition by deleting from the sum the $n = q$ term (which is $1/2q^2$), we have, after also reducing the right-hand side of (4.2.10) by $1/2q^2$,

(4.2.11) $$\sum_{\substack{n=1 \\ n \neq q}}^{\infty} \frac{1}{n(n+q)} = \frac{h(q)}{q} - \frac{1}{2q^2}.$$

Next, subtract (4.2.11) from (4.2.8). That is, write

(4.2.12) $$\sum_{\substack{n=1 \\ n \neq q}}^{\infty} \left\{ \frac{1}{n(n-q)} - \frac{1}{n(n+q)} \right\} = \left\{ \frac{2}{q^2} - \frac{h(q)}{q} \right\} - \left\{ \frac{h(q)}{q} - \frac{1}{2q^2} \right\}$$

$$= \frac{2}{q^2} + \frac{1}{2q^2} - 2\frac{h(q)}{q} = \frac{5}{2q^2} - 2\frac{h(q)}{q}.$$

Since

$$\frac{1}{n(n-q)} - \frac{1}{n(n+q)} = \frac{2q}{n(n^2 - q^2)},$$

then (4.2.12) becomes

$$2\sum_{\substack{n=1 \\ n \neq q}}^{\infty} \frac{q}{n(n^2 - q^2)} = \frac{5}{2q^2} - 2\frac{h(q)}{q},$$

which, if we divide through by $2q^2$ and then sum over all q, becomes

(4.2.13) $$\sum_{q=1}^{\infty} \sum_{\substack{n=1 \\ n \neq q}}^{\infty} \frac{1}{nq(n^2 - q^2)} = \sum_{q=1}^{\infty} \frac{5}{4q^4} - \sum_{q=1}^{\infty} \frac{h(q)}{q^3}.$$

The double sum in (4.2.13) vanishes for the same reason the double sum vanishes in (4.2.9), and we arrive at

$$(4.2.14) \qquad \sum_{q=1}^{\infty} \frac{h(q)}{q^3} = \frac{5}{4} \sum_{q=1}^{\infty} \frac{1}{q^4} = \frac{5}{4} \zeta(4)$$

and so, just like that, we have Euler's sum of (4.1.2).

To end this section, I claim that with even more simple algebra, we can derive a virtually endless sequence of similar expressions (all now called Euler sums). But I warn you: Once you get started doing this, it's like eating peanuts—it's really hard to stop! As an illustration of this claim, suppose we start with the sum

$$\sum_{q=2}^{\infty} \frac{h(q-1)}{q^m}$$

and then change index to $k = q - 1$ (and so $q = k + 1$). Doing that allows us to write

$$\sum_{q=2}^{\infty} \frac{h(q-1)}{q^m} = \sum_{k=1}^{\infty} \frac{h(k)}{(k+1)^m} = \sum_{q=1}^{\infty} \frac{h(q)}{(q+1)^m}.$$

From this we can write

$$\sum_{q=1}^{\infty} \frac{h(q)}{q^m} - \sum_{q=1}^{\infty} \frac{h(q)}{(q+1)^m} = \sum_{q=1}^{\infty} \frac{h(q)}{q^m} - \sum_{q=2}^{\infty} \frac{h(q-1)}{q^m}$$

$$= \left\{ h(1) + \sum_{q=2}^{\infty} \frac{h(q)}{q^m} \right\} - \sum_{q=2}^{\infty} \frac{h(q-1)}{q^m} = h(1) + \sum_{q=2}^{\infty} \frac{h(q) - h(q-1)}{q^m}.$$

Now, as $h(1) = 1$, and since

$$h(q) - h(q-1) = \left(1 + \frac{1}{2} + \cdots + \frac{1}{q} \right) - \left(1 + \frac{1}{2} + \cdots + \frac{1}{q-1} \right) = \frac{1}{q},$$

we therefore have

$$\sum_{q=1}^{\infty}\frac{h(q)}{q^m}-\sum_{q=1}^{\infty}\frac{h(q)}{(q+1)^m}=1+\sum_{q=2}^{\infty}\frac{\frac{1}{q}}{q^m}=1+\sum_{q=2}^{\infty}\frac{1}{q^{m+1}}$$

or

(4.2.15) $$\sum_{q=1}^{\infty}\frac{h(q)}{q^m}-\sum_{q=1}^{\infty}\frac{h(q)}{(q+1)^m}=\sum_{q=1}^{\infty}\frac{1}{q^{m+1}}=\zeta(m+1).$$

For example, suppose $m = 1$. Then (4.2.15) becomes

$$\sum_{q=1}^{\infty}\left\{\frac{h(q)}{q}-\frac{h(q)}{q+1}\right\}=\zeta(2).$$

That is, since

$$\frac{1}{q}-\frac{1}{q+1}=\frac{1}{q(q+1)}$$

then

(4.2.16) $$\sum_{q=1}^{\infty}\frac{h(q)}{q(q+1)}=\zeta(2).$$

We can check (4.2.16) with **euler1775.m** by simply changing the line ending in /q^2; to /(q*(q+1));, and by also changing the line s(1)=1; to s(1)=1/2;. This produces a value for the 10-millionth partial sum of 1.644932 ..., which compares pretty well with $\zeta(2)=\frac{\pi^2}{6}=1.644934\ldots$. And if $m = 2$ then (4.2.15) becomes

$$\sum_{q=1}^{\infty}\left\{\frac{h(q)}{q^2}-\frac{h(q)}{(q+1)^2}\right\}=\zeta(3).$$

So

$$\sum\nolimits_{q=1}^{\infty} \frac{h(q)}{(q+1)^2} = \sum\nolimits_{q=1}^{\infty} \frac{h(q)}{q^2} - \zeta(3)$$

or, recalling (4.1.1),

$$\sum\nolimits_{q=1}^{\infty} \frac{h(q)}{(q+1)^2} = 2\zeta(3) - \zeta(3)$$

and so

(4.2.17)
$$\sum\nolimits_{q=1}^{\infty} \frac{h(q)}{(q+1)^2} = \zeta(3).$$

We can check (4.2.17) with **euler1775.m** by simply changing the line ending in */q^2;* to */(q+1)^2;* as well as changing the line *s(1)=1;* to *s(1)=1/4;*. This produces a value for the 10-millionth partial sum of $1.202055\ldots$, which compares pretty well with $\zeta(3) = 1.202056\ldots$.

Euler sums can be exquisitely sensitive to seemingly minor changes. For example, what is

$$\sum\nolimits_{q=1}^{\infty} \frac{h(q)}{(q+2)^2} = ?$$

which looks similar to (4.2.17). As the following analysis shows, however, the answer is quite different from $\zeta(3)$. We start with (4.2.17) and write

$$\zeta(3) = \sum\nolimits_{q=1}^{\infty} \frac{h(q)}{(q+1)^2} = \sum\nolimits_{q=1}^{\infty} \frac{h(q-1)+\frac{1}{q}}{(q+1)^2} = \sum\nolimits_{q=1}^{\infty} \frac{h(q-1)}{(q+1)^2} + \sum\nolimits_{q=1}^{\infty} \frac{1}{q(q+1)^2}.$$

For the sums on the right change index to $q = k + 1$ ($k = q - 1$). Then

$$\sum_{q=1}^{\infty} \frac{h(q-1)}{(q+1)^2} = \sum_{k=0}^{\infty} \frac{h(k)}{(k+2)^2} = \sum_{k=1}^{\infty} \frac{h(k)}{(k+2)^2}$$

because $h(0) = 0$, and also

$$\sum_{q=1}^{\infty} \frac{1}{q(q+1)^2} = \sum_{k=0}^{\infty} \frac{1}{(k+1)(k+2)^2}.$$

Thus,

$$\zeta(3) = \sum_{k=1}^{\infty} \frac{h(k)}{(k+2)^2} + \sum_{k=0}^{\infty} \frac{1}{(k+1)(k+2)^2}$$

and so

(4.2.18) $$\sum_{q=1}^{\infty} \frac{h(q)}{(q+2)^2} = \zeta(3) - \sum_{k=0}^{\infty} \frac{1}{(k+1)(k+2)^2}.$$

We evaluate the sum on the right of (4.2.18) as follows. Change index to $q = k + 2$ (and so $q - 1 = k + 1$) which says

$$\sum_{k=0}^{\infty} \frac{1}{(k+1)(k+2)^2} = \sum_{q=2}^{\infty} \frac{1}{(q-1)q^2} = \sum_{q=2}^{\infty} \left\{ \frac{1}{q(q-1)} - \frac{1}{q^2} \right\}$$

$$= \sum_{q=2}^{\infty} \frac{1}{q(q-1)} - \sum_{q=2}^{\infty} \frac{1}{q^2}$$

$$= \sum_{q=2}^{\infty} \frac{1}{q(q-1)} - \left\{ \sum_{q=1}^{\infty} \frac{1}{q^2} - 1 \right\} = \sum_{q=2}^{\infty} \frac{1}{q(q-1)} - \{\zeta(2) - 1\}$$

$$= \sum_{q=2}^{\infty} \frac{1}{q(q-1)} - \zeta(2) + 1.$$

Now,

$$\sum_{q=2}^{\infty}\frac{1}{q(q-1)}=\sum_{q=2}^{\infty}\left\{\frac{1}{q-1}-\frac{1}{q}\right\}=\left\{1-\frac{1}{2}\right\}+\left\{\frac{1}{2}-\frac{1}{3}\right\}+\left\{\frac{1}{3}-\frac{1}{4}\right\}+\cdots=1$$

as the terms in the curly brackets telescope. So,

$$(4.2.19) \qquad \sum_{k=0}^{\infty}\frac{1}{(k+1)(k+2)^2}=2-\zeta(2)$$

and thus, at last, putting (4.2.19) into (4.2.18), we have the answer to our question:

$$(4.2.20) \qquad \sum_{q=1}^{\infty}\frac{h(q)}{(q+2)^2}=\zeta(3)+\zeta(2)-2.$$

Typing *zeta(3)+zeta(2)-2* into *MATLAB* returns a value of 0.846990.... We can check this result with **euler1775.m** by simply changing the line ending in */q^2;* to */(q+2)^2;*, as well as changing the line *s(1)=1;* to *s(1)=1/9;*. This produces a value for the ten millionth partial sum of 0.846989..., which compares pretty well with $\zeta(3)+\zeta(2)-2$.

Challenge Problem 4.2.1: After looking at (4.2.17), it's natural to next ask for the value of $\sum_{q=1}^{\infty}\frac{h(q)}{(q+1)^3}$. We can use **euler1775.m** to numerically estimate this sum by changing the line ending in */q^2;* to */(q+1)^3;*, as well as changing the line *s(1)=1;* to *s(1)=1/8;*. This produces a value for the 10-millionth partial sum of 0.2705808..., and you should check to see if this is consistent with your theoretical result. Hint: Start with (4.1.2) and remember the trick of writing $h(q)=h(q-1)+\frac{1}{q}$. Also, it will be helpful to recall that $\zeta(4)=\frac{\pi^4}{90}$ and that $h(0)=0$.

Challenge Problem 4.2.2: Show that $\int_0^1 x^{q-1}\ln(1-x)dx=-\frac{h(q)}{q}$. Hint: Start by showing $\ln(1-x)=-\int_0^x\frac{dt}{1-t}$, then put this into the original integral (thus getting a double integral), and then reverse the order of integration. We'll use the result in the final section of this

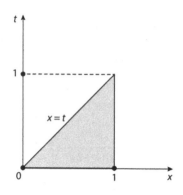

FIGURE 4.2.1.

The region of integration in the double integral.

chapter. *Special Note*: When you reverse the order of integration, you'll need to specify the limits of integration on x and t. To do that, you'll find it helpful to sketch the region of integration in the xt-plane. Since x varies from 0 to 1, and t varies from 0 to x, that region is the interior of the shaded triangle shown in Figure 4.2.1. If we do the integration of the reversed double integral as horizontal strips then, as t varies from 0 to 1, x varies from t to 1.

Why does $\sum_{q=1}^{\infty}\sum_{\substack{n=1\\n\neq q}}^{\infty}\frac{1}{qn(n-q)}=0$ in (4.2.9)? Let's assume this double sum has a finite value S. (Since both n and q are to the second power in the denominator, it is at least plausible that the sum doesn't diverge.) Now, think of the sum as a matrix of terms (I'll tell you the value of each term in just a moment) made of an infinity of rows, with each row having an infinity of terms (or, alternatively, as an infinity of columns, with each column having an infinity of terms). Suppose q is our row index, and n is our column index ($q = 1$ is the top row, and $n = 1$ is the left column). So, to start the summation, we set $q = 1$ and then let n run through all the terms in that row, that is, n runs from 2 to infinity (*not* 1 to infinity because of the $n \neq q$ constraint). The value of the terms in the first row is given by $\frac{1}{qn(n-q)}\big|_{q=1} = \frac{1}{n(n-1)}\big|_{n \neq 1}$, where n is the column index. Then we increase q by 1 to $q = 2$ and let n run through all the terms in the second row. The value of the terms in the second row is given by $\frac{1}{qn(n-q)}\big|_{q=2} = \frac{1}{2n(n-2)}\big|_{n \neq 2}$. Repeat this process over and over and over. When finally done (what being "done" means in this doubly infinite process has metaphysical issues to it that we'll ignore!), we'll have the sum S (*assumed*, remember, to be finite), given by

$$\sum_{\substack{\text{row index}}} \sum_{\substack{\text{column index} \\ \text{column index} \neq \text{row index}}} \frac{1}{(\text{row index})(\text{column index})(\text{column index} - \text{row index})}.$$

Next, let's do the double sum again, but now we'll add the terms in our matrix in a different order. Suppose that now q is the column index, and n is the row index. This will give us a sum of

$$\sum_{\substack{\text{column index}}} \sum_{\substack{\text{row index} \\ \text{row index} \neq \text{column index}}} \frac{1}{(\text{column index})(\text{row index})(\text{row index} - \text{column index})}$$

which is clearly the *negative* of the first summation (look at the right-most factor in the denominator of each of the two double sums). That is, we get $-S$. So, $S = -S$, and we conclude that $S = 0$, because zero is the only finite number equal to its negative. In Appendix 3 we'll show that S is indeed finite.

4.3 Euler's Double Sums

A general class of sums that Euler studied is defined by[1]

$$(4.3.1) \qquad S_{a,b} = \sum_{q=1}^{\infty} \frac{h^{(a)}(q)}{q^b}, \ h^{(a)}(q) = \sum_{k=1}^{q} \frac{1}{k^a}.$$

---

1. Note that $h^{(a)}(q) \neq h^a(q)$. That is, $\sum_{k=1}^{q} \frac{1}{k^a} \neq \{\sum_{k=1}^{q} \frac{1}{k}\}^a$.

If $a = 1$, then the special cases of $b = 2$ and $b = 3$ are (4.1.1) and (4.1.2), respectively. In the literature it is stated (without proof) that these sums obey the identity[2]

$$(4.3.2) \qquad S_{a,b} + S_{b,a} = \zeta(a)\zeta(b) + \zeta(a+b), \ a, b \geq 2.$$

In particular, if $a = b$, then (4.3.2) says

$$S_{a,a} + S_{a,a} = \zeta(a)\zeta(a) + \zeta(a+a) = 2S_{a,a}$$

and so

$$(4.3.3) \qquad S_{a,a} = \frac{1}{2}\{\zeta^2(a) + \zeta(2a)\}.$$

Is (4.3.3) correct?

To quickly see if it is worth further time and effort to explore that question, one possible approach is to just numerically evaluate both sides of (4.3.3) for some particular values of a (using, let's say, the values of q from one to 10 million). For example, if $a = 2$ then the right-hand side of (4.3.3) is

$$\frac{1}{2}\{\zeta^2(2) + \zeta(4)\} = \frac{1}{2}\left\{\left(\frac{\pi^2}{6}\right)^2 + \frac{\pi^4}{90}\right\} = \frac{1}{2}\left\{\frac{\pi^4}{36} + \frac{\pi^4}{90}\right\}$$

$$= \frac{7\pi^4}{360} = \frac{7}{4}\zeta(4) = 1.89406\ldots.$$

2. Philippe Flajolet and Bruno Salvy, "Euler Sums and Contour Integral Representations," *Experimental Mathematics*, vol. 7 (no. 1), 1998, pp. 15–35.

The left-hand side of (4.3.3) is

$$\sum_{q=1}^{\infty}\left\{\frac{\sum_{k=1}^{q}\dfrac{1}{k^2}}{q^2}\right\} = \sum_{q=1}^{\infty}\frac{1}{q^2}\left(1+\frac{1}{2^2}+\frac{1}{3^2}+\cdots+\frac{1}{q^2}\right)$$

and we'll need a computer to estimate this expression (using $q = 1$ to 10,000,000). The *MATLAB* code **checksum.m** in the next box does the job. When executed, **checksum.m** produced the estimate $S_{2,2} = 1.89406\ldots$, which is pretty close to the claimed theoretical value. I think, therefore, that most analysts would conclude that it is worth the effort to try to establish (4.3.3) for any integer $a \geq 2$, not just for $a = 2$. Here's how to do that.

```
%checksum.m
q=1;h(q)=1;
for q=2:10000000
    h(q)=h(q-1)+1/q^2;
end
S=0;
for q=1:10000000
    S=S+h(q)/q^2;
end
S
```

If we explicitly write out what $S_{a,a}$ is, we get

$$S_{a,a} = \sum_{q=1}^{\infty}\frac{\sum_{k=1}^{q}\dfrac{1}{k^a}}{q^a} = \sum_{q=1}^{\infty}\frac{1}{q^a}\left\{\frac{1}{1^a}+\frac{1}{2^a}+\frac{1}{3^a}+\cdots+\frac{1}{q^a}\right\}$$

$$= \sum_{q=1}^{\infty}\frac{1}{q^a}\left[\left\{\frac{1}{1^a}+\frac{1}{2^a}+\frac{1}{3^a}+\cdots+\frac{1}{q^a}\right\}+\left\{\frac{1}{(q+1)^a}+\frac{1}{(q+2)^a}+\cdots\right\}\right.$$

$$-\left\{\frac{1}{(q+1)^a}+\frac{1}{(q+2)^a}+\cdots\right\}\Bigg]$$

$$=\sum_{q=1}^{\infty}\frac{1}{q^a}\left\{\frac{1}{1^a}+\frac{1}{2^a}+\frac{1}{3^a}+\cdots\right\}-\sum_{q=1}^{\infty}\frac{1}{q^a}\left\{\frac{1}{(q+1)^a}+\frac{1}{(q+2)^a}+\cdots\right\}$$

$$=\sum_{q=1}^{\infty}\frac{1}{q^a}\zeta(a)-\sum_{q=1}^{\infty}\frac{1}{q^a}\left\{\frac{1}{(q+1)^a}+\frac{1}{(q+2)^a}+\cdots\right\}$$

$$=\zeta(a)\sum_{q=1}^{\infty}\frac{1}{q^a}-\sum_{q=1}^{\infty}\frac{1}{q^a}\left\{\frac{1}{(q+1)^a}+\frac{1}{(q+2)^a}+\cdots\right\}$$

$$=\zeta^2(a)-\sum_{q=1}^{\infty}\frac{1}{q^a}\left\{\frac{1}{(q+1)^a}+\frac{1}{(q+2)^a}+\cdots\right\}$$

$$=\left\{\zeta^2(a)-\frac{1}{2}\zeta^2(a)\right\}+\frac{1}{2}\zeta^2(a)-\sum_{q=1}^{\infty}\frac{1}{q^a}\left\{\frac{1}{(q+1)^a}+\frac{1}{(q+2)^a}+\cdots\right\}$$

or

$$(4.3.4)\quad S_{a,a}=\frac{1}{2}\zeta^2(a)+\left[\frac{1}{2}\zeta^2(a)-\sum_{q=1}^{\infty}\frac{1}{q^a}\left\{\frac{1}{(q+1)^a}+\frac{1}{(q+2)^a}+\cdots\right\}\right].$$

Now, concentrate on the terms on the right in the square brackets of (4.3.4). That is, on

$$\frac{1}{2}\zeta^2(a)-\sum_{q=1}^{\infty}\frac{1}{q^a}\left\{\frac{1}{(q+1)^a}+\frac{1}{(q+2)^a}+\cdots\right\}$$

$$=\frac{1}{2}\left\{\frac{1}{1^a}+\frac{1}{2^a}+\frac{1}{3^a}+\cdots\right\}\left\{\frac{1}{1^a}+\frac{1}{2^a}+\frac{1}{3^a}+\cdots\right\}$$

$$-\sum_{q=1}^{\infty}\frac{1}{q^a}\left\{\frac{1}{(q+1)^a}+\frac{1}{(q+2)^a}+\cdots\right\}.$$

When the expressions in the first two pairs of curly brackets are multiplied, we'll get terms that are the squares of each term, plus all the cross-products of those terms as well. That is,

$$\frac{1}{2}\zeta^2(a) - \sum_{q=1}^{\infty} \frac{1}{q^a} \left\{ \frac{1}{(q+1)^a} + \frac{1}{(q+2)^a} + \cdots \right\}$$

$$= \frac{1}{2}\left\{ \left[\frac{1}{1^{2a}} + \frac{1}{2^{2a}} + \frac{1}{3^{2a}} + \cdots \right] + \text{cross-products} \right\}$$

$$- \sum_{q=1}^{\infty} \frac{1}{q^a} \left\{ \frac{1}{(q+1)^a} + \frac{1}{(q+2)^a} + \cdots \right\}$$

or

$$\frac{1}{2}\zeta^2(a) - \sum_{q=1}^{\infty} \frac{1}{q^a} \left\{ \frac{1}{(q+1)^a} + \frac{1}{(q+2)^a} + \cdots \right\}$$

$$= \frac{1}{2}\left[\zeta(2a) + \text{cross-products} \right]$$

$$- \sum_{q=1}^{\infty} \frac{1}{q^a} \left\{ \frac{1}{(q+1)^a} + \frac{1}{(q+2)^a} + \cdots \right\}.$$

Thus

$$(4.3.5) \qquad S_{a,a} = \frac{1}{2}\zeta^2(a) + \frac{1}{2}\zeta(2a) + \frac{1}{2}\left[\text{cross-products} \right]$$

$$- \sum_{q=1}^{\infty} \frac{1}{q^a} \left\{ \frac{1}{(q+1)^a} + \frac{1}{(q+2)^a} + \cdots \right\}.$$

To see what the cross-product terms are like, it's helpful to work through a specific example. Suppose we have the product $(A + B + C$

$+ D)(A + B + C + D)$. If you multiply this out, you get $\{A^2 + B^2 + C^2 + D^2\} + 2\{AB + AC + AD + BC + BD + CD\}$. The first curly brackets of the product are the squared terms, while the second curly brackets show that the cross-products are formed by multiplying each term in one of the original factors by each of the terms *to its right*, and then by 2. Applying this observation to our problem, we find

$$\frac{1}{2}[\text{cross-products}] - \sum_{q=1}^{\infty} \frac{1}{q^a}\left\{\frac{1}{(q+1)^a} + \frac{1}{(q+2)^a} + \cdots\right\} = 0,$$

which reduces (4.3.5) to the claim of (4.3.3).

Challenge Problem 4.3.1: See if you can establish the identity $\sum_{q=1}^{\infty} x^q h(q) = -\frac{\ln(1-x)}{1-x}$, $-1 < x \leq 1$. Hint: Recalling (1.3.5) will be helpful.

4.4 Euler Sums after Euler

An interesting class of exotic Euler sums (that I don't believe Euler himself ever actually investigated) has $h^2(q)$ in the numerator. For example, what is

$$\sum_{q=1}^{\infty} \frac{h^2(q)}{q(q+1)} = ?$$

We can answer this question as follows, starting by writing

(4.4.1) $$\sum_{q=1}^{\infty} \frac{h^2(q)}{q(q+1)} = \sum_{q=1}^{\infty} \frac{h^2(q)}{q} - \sum_{q=1}^{\infty} \frac{h^2(q)}{q+1}.$$

In the last sum on the right-hand side of (4.4.1), change the index to $k = q + 1$ ($q = k - 1$). Then

$$\sum_{q=1}^{\infty} \frac{h^2(q)}{q+1} = \sum_{k=2}^{\infty} \frac{h^2(k-1)}{k}$$

and so we have

$$\sum_{q=1}^{\infty} \frac{h^2(q)}{q(q+1)} = \sum_{q=1}^{\infty} \frac{h^2(q)}{q} - \sum_{q=2}^{\infty} \frac{h^2(q-1)}{q}$$

$$= \sum_{q=1}^{\infty} \frac{h^2(q)}{q} - \left\{ \sum_{q=1}^{\infty} \frac{h^2(q-1)}{q} - h^2(0) \right\}$$

or, as $h(0) = 0$,

(4.4.2) $$\sum_{q=1}^{\infty} \frac{h^2(q)}{q(q+1)} = \sum_{q=1}^{\infty} \frac{h^2(q)-h^2(q-1)}{q}.$$

Now,

$$h^2(q) - h^2(q-1) = \{h(q)+h(q-1)\}\{h(q)-h(q-1)\}$$

and, as

$$h(q) - h(q-1) = \left\{1+\frac{1}{2}+\frac{1}{3}+\cdots+\frac{1}{q}\right\} - \left\{1+\frac{1}{2}+\frac{1}{3}+\cdots+\frac{1}{q-1}\right\} = \frac{1}{q}$$

then (4.4.2) becomes

$$\sum_{q=1}^{\infty} \frac{h^2(q)}{q(q+1)} = \sum_{q=1}^{\infty} \frac{\frac{1}{q}\{h(q)+h(q-1)\}}{q}$$

or

(4.4.3) $$\sum_{q=1}^{\infty} \frac{h^2(q)}{q(q+1)} = \sum_{q=1}^{\infty} \frac{h(q)}{q^2} + \sum_{q=1}^{\infty} \frac{h(q-1)}{q^2}.$$

Since

$$h(q-1) = h(q) - \frac{1}{q}$$

we see that (4.4.3) becomes

$$\sum_{q=1}^{\infty}\frac{h^2(q)}{q(q+1)}=\sum_{q=1}^{\infty}\frac{h(q)}{q^2}+\sum_{q=1}^{\infty}\frac{h(q)}{q^2}-\sum_{q=1}^{\infty}\frac{1}{q^3}$$

or, recalling (4.1.1),

$$\sum_{q=1}^{\infty}\frac{h^2(q)}{q(q+1)}=2\sum_{q=1}^{\infty}\frac{h(q)}{q^2}-\sum_{q=1}^{\infty}\frac{1}{q^3}=2\{2\zeta(3)\}-\zeta(3)$$

and so

(4.4.4) $$\sum_{q=1}^{\infty}\frac{h^2(q)}{q(q+1)}=3\zeta(3).$$

We can check (4.4.4) with a slightly modified **checksum.m**, a code that I'll call **numsquared.m** (see the following box). When run, **numsquared.m** produced the estimate for the sum's value of 3.60613 ..., which compares nicely with the value of $3\zeta(3) = 3.60617 \ldots$.

```
%numsquared.m
q=1;h(q)=1;
for q=2:10000000
    h(q)=h(q-1)+1/q;
end
for q=1:10000000
    n(q)=h(q)^2;
end
S=0;
for q=1:10000000
    S=S+n(q)/(q*(q+1));
end
S
```

To end this chapter on a dramatic note, we'll next do two really spectacular Euler sums. To start, I'll show you that

$$\sum_{q=1}^{\infty} \frac{h^2(q)}{q^2} = \frac{17}{4}\zeta(4),$$

a result that has an especially interesting backstory that I'll tell you more about in the challenge problem that ends the book. The sum looks a lot like the sum in (4.4.4), but it's pretty clear that $3\zeta(3)$ is quite different in form from $\frac{17}{4}\zeta(4)$. The method I'm about to show you appeared[3] in 2015, long after the answer had been found by other means, and so this is a nice example of how a math problem, even if solved, can still offer a ripe opportunity for finding a new solution. The analysis will be one of the longer treatments in the book, consisting of a sequence of seemingly unrelated calculations. Each of those calculations is not particularly difficult—all are within the grasp of high school algebra (and of AP-calculus, if doing double integrals is acceptable), but taken together, the calculations are sufficient to suddenly solve our problem. So, we start.

Let $a_1, a_2, a_3, \ldots$ and $b_1, b_2, b_3, \ldots$ denote two sequences of arbitrary length of real numbers. If

$$A_q = \sum_{k=1}^{q} a_k, \ A_0 = 0,$$

that is, if A_q is the sum of the first q numbers in the a-sequence, then the claim is

(4.4.5) $\sum_{k=1}^{q} a_k b_k = A_q b_{q+1} + \sum_{k=1}^{q} A_k \left(b_k - b_{k+1}\right),$

3. My discussion here is an elaboration of the analysis given by Cornel Ioan Vălean and Ovidiu Furdui in their paper, "Reviving the Quadratic Series of Au-Yeung," *Journal of Classical Analysis*, vol. 6 (no. 2), 2015, pp. 113–118. The reference to Au-Yeung will be explained in the challenge problem at the end of this chapter.

a result called Abel's summation formula (see Challenge Problem 2.1.2 for more on Abel). We can derive (4.4.5) as follows, starting with the observation

$$a_k = (a_1 + a_2 + \cdots + a_k) - (a_1 + a_2 + \cdots + a_{k-1}) = A_k - A_{k-1}.$$

So

$$\sum_{k=1}^{q} a_k b_k = \sum_{k=1}^{q} (A_k - A_{k-1}) b_k = \sum_{k=1}^{q} A_k b_k - \sum_{k=1}^{q} A_{k-1} b_k.$$

Since $A_0 = 0$, the last sum becomes

$$\sum_{k=1}^{q} A_{k-1} b_k = \sum_{k=2}^{q} A_{k-1} b_k$$

and so

$$\sum_{k=1}^{q} a_k b_k = \sum_{k=1}^{q} A_k b_k - \sum_{k=2}^{q} A_{k-1} b_k.$$

In the last sum, let $j = k - 1$, and so

$$\sum_{k=2}^{q} A_{k-1} b_k = \sum_{j=1}^{q-1} A_j b_{j+1} = \sum_{k=1}^{q-1} A_k b_{k+1}.$$

Thus,

$$\sum_{k=1}^{q} a_k b_k = \sum_{k=1}^{q} A_k b_k - \sum_{k=1}^{q-1} A_k b_{k+1}$$

or, as

$$\sum_{k=1}^{q-1} A_k b_{k+1} = \sum_{k=1}^{q} A_k b_{k+1} - A_q b_{q+1}$$

we have

$$\sum_{k=1}^{q} a_k b_k = A_q b_{q+1} + \sum_{k=1}^{q} A_k b_k - \sum_{k=1}^{q} A_k b_{k+1},$$

which is (4.4.5).

Putting the Abel summation formula temporarily to the side (but don't push it too far away, as we'll be returning to it soon), let's next do the integral

$$\int_0^1 x^{q-1} \ln^2(1-x)dx,$$

which may remind you of the integral in Challenge Problem 4.2.2. We start our evaluation by observing that

(4.4.6) $$\ln^2(1-x) = \int_0^x -2\frac{\ln(1-t)}{1-t}dt.$$

To see that this is so, change variable in the integral to $u = 1 - t$ (and so $dt = -du$). Then,

$$\int_0^x -2\frac{\ln(1-t)}{1-t}dt = \int_1^{1-x} -2\frac{\ln(u)}{u}(-du) = -2\int_{1-x}^1 \frac{\ln(u)}{u}du$$

$$= -2\left\{\frac{1}{2}\ln^2(u)\right\}\Big|_{1-x}^1 = \ln^2(1-x)$$

as claimed.

Inserting (4.4.6) into our original integral, we have

$$\int_0^1 x^{q-1} \ln^2(1-x)dx = \int_0^1 x^{q-1}\left\{\int_0^x -2\frac{\ln(1-t)}{1-t}dt\right\}dx$$

$$= -2\int_0^1 \frac{\ln(1-t)}{1-t}\left\{\int_t^1 x^{q-1}dx\right\}dt$$

where the order of integration has been reversed (an operation a mathematician would want to spend some time in justifying) and we've used the Special Note in Challenge Problem 4.2.2 to get the limits on the x-integration. Continuing,

$$\int_0^1 x^{q-1} \ln^2(1-x)dx = -2\int_0^1 \frac{\ln(1-t)}{1-t}\left\{\frac{x^q}{q}\right\}\Big|_t^1 dt$$

$$= -\frac{2}{q}\int_0^1 \frac{\ln(1-t)}{1-t}(1-t^q)dt.$$

As shown in the solution to Challenge Problem 4.2.2,

$$\frac{1-t^q}{1-t} = 1+t+t^2+\cdots+t^{q-1}$$

and so

(4.4.7) $$\int_0^1 x^{q-1}\ln^2(1-x)dx = -\frac{2}{q}\int_0^1 (1+t+t^2+\cdots+t^{q-1})\ln(1-t)dt.$$

Now, recall what you showed (you did, didn't you?) in Challenge Problem 4.2.2:

$$\int_0^1 t^{n-1}\ln(1-t)dt = -\frac{h(n)}{n}$$

and we see that the right-hand side of (4.4.7) is the sum of numerous such integrals, one integral for each value of n as n runs through the integers 1 to q. Thus,

$$\int_0^1 x^{q-1}\ln^2(1-x)dx = -\frac{2}{q}\left\{-\frac{h(1)}{1}-\frac{h(2)}{2}-\cdots-\frac{h(q)}{q}\right\}$$

or

(4.4.8) $$\int_0^1 x^{q-1} \ln^2(1-x)\,dx = \frac{2}{q}\sum_{k=1}^q \frac{h(k)}{k}.$$

We can evaluate the sum in (4.4.8) using Abel's summation formula of (4.4.5). In that formula, set $a_k = \frac{1}{k}$ and $b_k = h(k)$. Then, as

$$A_q = 1 + \frac{1}{2} + \frac{1}{3} + \cdots + \frac{1}{q} = h(q)$$

$$A_k = 1 + \frac{1}{2} + \frac{1}{3} + \cdots + \frac{1}{k} = h(k)$$

and since

$$b_k - b_{k+1} = h(k) - h(k+1) = \left(1 + \frac{1}{2} + \frac{1}{3} + \cdots + \frac{1}{k}\right)$$

$$- \left(1 + \frac{1}{2} + \frac{1}{3} + \cdots + \frac{1}{k} + \frac{1}{k+1}\right) = -\frac{1}{k+1},$$

we see that (4.4.5) becomes

$$\sum_{k=1}^q \frac{h(k)}{k} = h(q)h(q+1) - \sum_{k=1}^q \frac{h(k)}{k+1}$$

$$= h(q)h(q+1) - \sum_{k=1}^q \frac{h(k+1) - \dfrac{1}{k+1}}{k+1}$$

$$= h(q)h(q+1) - \sum_{k=1}^q \frac{h(k+1)}{k+1} + \sum_{k=1}^q \frac{1}{(k+1)^2}.$$

In the last two sums, change the index to $j = k + 1$, and so

$$\sum_{k=1}^q \frac{h(k)}{k} = h(q)h(q+1) - \sum_{j=2}^{q+1} \frac{h(j)}{j} + \sum_{j=2}^{q+1} \frac{1}{j^2}$$

$$= h(q)h(q+1) - \left\{ \sum_{j=1}^{q} \frac{h(j)}{j} - h(1) + \frac{h(q+1)}{q+1} \right\}$$

$$+ \left\{ \sum_{j=1}^{q} \frac{1}{j^2} - 1 + \frac{1}{(q+1)^2} \right\}$$

or, as $h(1) = 1$, this becomes

(4.4.9) $$\sum_{k=1}^{q} \frac{h(k)}{k} = h(q)h(q+1) - \sum_{j=1}^{q} \frac{h(j)}{j}$$

$$-\frac{h(q+1)}{q+1} + \sum_{j=1}^{q} \frac{1}{j^2} + \frac{1}{(q+1)^2}.$$

We can greatly simplify (4.4.9) by noticing that

$$h(q)h(q+1) = h(q)\left\{ h(q) + \frac{1}{q+1} \right\} = h^2(q) + \frac{h(q)}{q+1}.$$

Using this in (4.4.9), it becomes

$$\sum_{k=1}^{q} \frac{h(k)}{k} = h^2(q) + \frac{h(q)}{q+1} - \sum_{j=1}^{q} \frac{h(j)}{j} - \frac{h(q+1)}{q+1} + \sum_{j=1}^{q} \frac{1}{j^2} + \frac{1}{(q+1)^2}$$

$$= h^2(q) - \sum_{j=1}^{q} \frac{h(j)}{j} + \sum_{j=1}^{q} \frac{1}{j^2} + \frac{1}{(q+1)^2} + \frac{h(q) - h(q+1)}{q+1}$$

$$= h^2(q) - \sum_{j=1}^{q} \frac{h(j)}{j} + \sum_{j=1}^{q} \frac{1}{j^2} + \frac{1}{(q+1)^2} + \frac{-\dfrac{1}{q+1}}{q+1}$$

or, as the last two terms cancel, we have

$$\sum_{k=1}^{q} \frac{h(k)}{k} = h^2(q) - \sum_{j=1}^{q} \frac{h(j)}{j} + \left(1 + \frac{1}{2^2} + \frac{1}{3^2} + \cdots + \frac{1}{q^2} \right).$$

Noticing that

$$\sum_{k=1}^{q} \frac{h(k)}{k} = \sum_{j=1}^{q} \frac{h(j)}{j}$$

we arrive at

$$(4.4.10) \quad h^2(q) + \left(1 + \frac{1}{2^2} + \frac{1}{3^2} + \cdots + \frac{1}{q^2}\right) = 2\sum_{k=1}^{q} \frac{h(k)}{k}.$$

Okay, I know this has been a long trek, and you are by now almost certainly wondering where in the heck we are going with all this, but hang in there for just a moment more. We are almost done!

Looking back at (4.4.8), we see that

$$q\int_0^1 x^{q-1} \ln^2(1-x)dx = 2\sum_{k=1}^{q} \frac{h(k)}{k},$$

which, combined with (4.4.10), says

$$q\int_0^1 x^{q-1} \ln^2(1-x)dx = h^2(q) + \left(1 + \frac{1}{2^2} + \frac{1}{3^2} + \cdots + \frac{1}{q^2}\right).$$

So, dividing through by q^2 and then summing over all q, we have

$$(4.4.11) \quad \sum_{q=1}^{\infty} \frac{1}{q} \int_0^1 x^{q-1} \ln^2(1-x)dx = \sum_{q=1}^{\infty} \frac{h^2(q)}{q^2}$$

$$+ \sum_{q=1}^{\infty} \frac{1}{q^2}\left(1 + \frac{1}{2^2} + \frac{1}{3^2} + \cdots + \frac{1}{q^2}\right).$$

The right-most sum of (4.4.11) is one we did in the previous section (the $a = 2$ case of (4.3.3)); we found that sum to be $\frac{7}{4}\zeta(4)$. So, reversing the order of summation and integration on the left-hand side of (4.4.11), we have

$$\int_0^1 \sum_{q=1}^{\infty} \frac{x^{q-1}}{q} \ln^2(1-x)dx = \sum_{q=1}^{\infty} \frac{h^2(q)}{q^2} + \frac{7}{4}\zeta(4)$$

or

$$(4.4.12) \quad \sum_{q=1}^{\infty} \frac{h^2(q)}{q^2} = \int_0^1 \ln^2(1-x)\sum_{q=1}^{\infty} \frac{x^{q-1}}{q}dx - \frac{7}{4}\zeta(4).$$

To do the integral in (4.4.12), recall the result we derived immediately after (2.1.12):

$$\frac{\ln(1-x)}{x} = -1 - \frac{1}{2}x - \frac{1}{3}x^2 - \frac{1}{4}x^3 - \frac{1}{5}x^4 - \cdots = -\sum_{q=1}^{\infty}\frac{x^{q-1}}{q}.$$

Using this in (4.4.12), we arrive at

$$(4.4.13) \quad \sum_{q=1}^{\infty}\frac{h^2(q)}{q^2} = -\int_0^1 \frac{\ln^3(1-x)}{x}dx - \frac{7}{4}\zeta(4).$$

Changing variable in the integral to $y = 1 - x$ ($dx = -dy$), we have

$$-\int_0^1 \frac{\ln^3(1-x)}{x}dx = -\int_1^0 \frac{\ln^3(y)}{1-y}(-dy) = -\int_0^1 \frac{\ln^3(y)}{1-y}dy$$

$$= -\int_0^1 \ln^3(y)\{1 + y + y^2 + y^3 + \cdots\}dy = -\int_0^1 \ln^3(y)\sum_{k=0}^{\infty}y^k dy$$

$$= -\sum_{k=0}^{\infty}\int_0^1 y^k \ln^3(y)dy$$

where, once again, I've assumed we can reverse the order of summation and integration. The integral is easy to do[4]—in fact, we've already done it, back in (1.6.13):

4. This particular integral can also be easily done with three successive integrations by parts, as hinted in the box near the end of Section 1.6. A generalization is also easy to do using the gamma function, an approach I won't repeat here, as you can find all the details in the second edition of my *Inside Interesting Integrals*, Springer 2020, pp. 179, 228, and 467: $\int_0^1 x^m \ln^n(x)dx = (-1)^n \frac{n!}{(m+1)^{n+1}}$.

$$\int_0^1 y^k \ln^3(y)\,dy = -\frac{6}{(k+1)^4}$$

and so

$$-\int_0^1 \frac{\ln^3(1-x)}{x}\,dx = 6\sum_{k=0}^{\infty} \frac{1}{(k+1)^4} = 6\left(\frac{1}{1^4}+\frac{1}{2^4}+\frac{1}{3^4}+\cdots\right) = 6\zeta(4).$$

Thus, putting this into (4.4.13), we have

$$(4.4.14) \quad \sum_{q=1}^{\infty} \frac{h^2(q)}{q^2} = 6\zeta(4) - \frac{7}{4}\zeta(4) = \frac{17}{4}\zeta(4) = \frac{17\pi^4}{360} = 4.59987\ldots$$

and we are—finally!—done. We can check (4.4.14) with **numsquared.m** by simply changing the line ending in */(q*(q+1));* to */q^2;*. The code produces a value for the 10-millionth partial sum of the sum on the left-hand side of (4.4.14) of 4.59984 . . . , which is in pretty good agreement with our theoretical result.

For our final example of Euler sum calculation, we'll do something we haven't done before—an *alternating* sum. Specifically, let's calculate the value of

$$\sum_{q=1}^{\infty} (-1)^{q-1}\frac{h(q)}{q^2} = ?$$

This looks almost like the Euler sum (4.1.1) that opens this chapter, but with the difference that the terms alternate in sign. As in (4.1.1), we'll find $\zeta(3)$ appears in the result, but now with a new coefficient. To do this sum, I'll first gather together some preliminary results we'll need along the way. The first one is particularly easy, as we've already done it.

If you look back once more at the box at the end of Section 1.6, you'll see there that we derived the integral

$$\int_0^1 v^n \ln^2(v)dv = \frac{2}{(n+1)^3}$$

and so (with a trivial change in notation)

(4.4.15) $$\int_0^1 x^a \ln^2(x)dx = \frac{2}{(a+1)^3}.$$

Next, recall from (4.4.6) that

$$\ln^2(1-x) = \int_0^x -2\frac{\ln(1-t)}{1-t}dt.$$

From (2.1.12) we have

$$\ln(1-x) = -x - \frac{1}{2}x^2 - \frac{1}{3}x^3 - \frac{1}{4}x^4 - \cdots$$

and so

$$\ln^2(1-x) = \int_0^x -2\frac{-t - \frac{1}{2}t^2 - \frac{1}{3}t^3 - \frac{1}{4}t^4 - \cdots}{1-t}dt$$

$$= 2\int_0^x \left(t + \frac{1}{2}t^2 + \frac{1}{3}t^3 + \frac{1}{4}t^4 + \cdots\right)(1 + t + t^2 + t^3 + t^4 + \cdots)dt$$

$$= 2\int_0^x \left\{\left(t + \frac{1}{2}t^2 + \frac{1}{3}t^3 + \frac{1}{4}t^4 + \cdots\right) + \left(t^2 + \frac{1}{2}t^3 + \frac{1}{3}t^4 + \frac{1}{4}t^5 + \cdots\right)\right.$$

$$\left. + \left(t^3 + \frac{1}{2}t^4 + \frac{1}{3}t^5 + \cdots\right) + \cdots\right\}dt$$

$$= 2\int_0^x \left\{t + \left(1 + \frac{1}{2}\right)t^2 + \left(1 + \frac{1}{2} + \frac{1}{3}\right)t^3 + \left(1 + \frac{1}{2} + \frac{1}{3} + \frac{1}{4}\right)t^4 + \cdots\right\}dt$$

$$= 2\left[\frac{1}{2}t^2 + \frac{1+\frac{1}{2}}{3}t^3 + \frac{1+\frac{1}{2}+\frac{1}{3}}{4}t^4 + \frac{1+\frac{1}{2}+\frac{1}{3}+\frac{1}{4}}{5}t^5 + \cdots\right]\Big|_0^x$$

$$= 2\left(\frac{1}{2}x^2 + \frac{1+\frac{1}{2}}{3}x^3 + \frac{1+\frac{1}{2}+\frac{1}{3}}{4}x^4 + \frac{1+\frac{1}{2}+\frac{1}{3}+\frac{1}{4}}{5}x^5 + \cdots\right)$$

and so we arrive at

$$\ln^2(1-x) = 2\sum_{q=1}^{\infty} x^{q+1} \frac{h(q)}{q+1}.$$

Or, writing $-x$ for x, and since $(-1)^{q+1} = (-1)^{q-1}$,

(4.4.16) $$\ln^2(1+x) = 2\sum_{q=1}^{\infty}(-1)^{q-1}x^{q+1}\frac{h(q)}{q+1}.$$

For our final preliminary calculation, we'll need to know the value of the integral

$$\int_0^1 \frac{\ln^2(1+x)}{x}dx = ?$$

As the starting point in doing this integral, you should have no difficulty in convincing yourself of the truth of the algebraic identity

$$B^2 = \frac{(A+B)^2 + (A-B)^2 - 2A^2}{2}.$$

So, if we define $A = \ln(1-x)$ and $B = \ln(1+x)$, it then immediately follows that

$$\ln^2(1+x) = \frac{\{\ln(1-x)+\ln(1+x)\}^2 + \{\ln(1-x)-\ln(1+x)\}^2 - 2\ln^2(1-x)}{2}$$

or if we divide through by x and integrate from 0 to 1,

(4.4.17)
$$\int_0^1 \frac{\ln^2(1+x)}{x}\,dx = \frac{1}{2}\int_0^1 \frac{\ln^2(1-x^2)}{x}\,dx$$

$$+ \frac{1}{2}\int_0^1 \frac{\ln^2\left(\dfrac{1-x}{1+x}\right)}{x}\,dx - \int_0^1 \frac{\ln^2(1-x)}{x}\,dx.$$

Now consider in turn each of the integrals on the right in (4.4.17).

For the first integral, make the change of variable $y = 1 - x^2$ (and so $dx = -\frac{dy}{2x}$). Then,

$$\int_0^1 \frac{\ln^2(1-x^2)}{x}\,dx = \int_1^0 \int_0^1 \frac{\ln^2(y)}{x}\left(-\frac{dy}{2x}\right) = \frac{1}{2}\int_0^1 \frac{\ln^2(y)}{x^2}\,dy = \frac{1}{2}\int_0^1 \frac{\ln^2(y)}{1-y}\,dy$$

$$= \frac{1}{2}\int_0^1 \{1 + y + y^2 + y^3 + \cdots\}\ln^2(y)\,dy = \frac{1}{2}\int_0^1 \sum_{k=1}^{\infty} y^{k-1}\ln^2(y)\,dy$$

$$= \frac{1}{2}\sum_{k=1}^{\infty}\int_0^1 x^{k-1}\ln^2(x)\,dx.$$

Recalling (4.4.15), we have, on setting $a = k - 1$, that

$$\int_0^1 \frac{\ln^2(1-x^2)}{x}\,dx = \frac{1}{2}\sum_{k=1}^{\infty}\frac{2}{k^3}$$

or

(4.4.18)
$$\int_0^1 \frac{\ln^2(1-x^2)}{x}\,dx = \zeta(3).$$

For the second integral on the right in (4.4.17), make the change of variable $y = \frac{1-x}{1+x}$ (and so $\frac{dy}{dx} = -\frac{2}{(1+x)^2}$ or $dx = -\frac{(1+x)^2}{2}\,dy$). Now, notice that

$$1 - y^2 = 1 - \left(\frac{1-x}{1+x}\right)^2 = \frac{4x}{(1+x)^2}$$

and so

$$(1+x)^2 = \frac{4x}{1-y^2}.$$

Thus,

$$dx = -\frac{4x}{2(1-y^2)}dy = -\frac{2x}{(1-y^2)}dy$$

and therefore

$$\int_0^1 \frac{\ln^2\left(\dfrac{1-x}{1+x}\right)}{x}dx = \int_1^0 \frac{\ln^2(y)}{x}\left(-\frac{2x}{(1-y^2)}dy\right)$$

$$= 2\int_0^1 \frac{\ln^2(y)}{1-y^2}dy$$

$$= 2\int_0^1 \{1+y^2+y^4+y^6+\cdots\}\ln^2(y)dy$$

$$= 2\int_0^1 \sum_{k=1}^{\infty} y^{2(k-1)}\ln^2(y)dy$$

$$= 2\sum_{k=1}^{\infty}\int_0^1 x^{2(k-1)}\ln^2(x)dx.$$

Again recalling (4.4.15), with a now set to $2(k-1) = 2k-2$, we have

$$\int_0^1 \frac{\ln^2\left(\dfrac{1-x}{1+x}\right)}{x}dx = 2\sum_{k=1}^{\infty}\frac{2}{(2k-1)^3} = 4\sum_{k=1}^{\infty}\frac{1}{(2k-1)^3}.$$

This last sum, that of the reciprocals cubed of the *odd* positive integers, is equal to

$$\left(\frac{1}{1^3}+\frac{1}{2^3}+\frac{1}{3^3}+\frac{1}{4^3}+\frac{1}{5^3}+\cdots\right)-\left(\frac{1}{2^3}+\frac{1}{4^3}+\frac{1}{6^3}+\cdots\right)$$

$$=\zeta(3)-\frac{1}{2^3}\left(\frac{1}{1^3}+\frac{1}{2^3}+\frac{1}{3^3}+\cdots\right)$$

$$=\zeta(3)-\frac{1}{8}\zeta(3).$$

That is,

$$\sum_{k=1}^{\infty}\frac{1}{(2k-1)^3}=\frac{7}{8}\zeta(3)$$

and so

(4.4.19)
$$\int_0^1\frac{\ln^2\left(\frac{1-x}{1+x}\right)}{x}dx=\frac{7}{2}\zeta(3).$$

Finally, for the third, right-most integral in (4.4.17), I'll let you verify that making the change of variable $x = y^2$ quickly leads to the result

(4.4.20)
$$\int_0^1\frac{\ln^2(1-x)}{x}dx=2\zeta(3).$$

With our results of (4.4.18), (4.4.19), and (4.4.20) in hand, we can now plug them into (4.4.17) to compute

$$\int_0^1\frac{\ln^2(1+x)}{x}dx=\frac{1}{2}\{\zeta(3)\}+\frac{1}{2}\left\{\frac{7}{2}\zeta(3)\right\}-2\zeta(3)$$

and arrive at

(4.4.21)
$$\int_0^1 \frac{\ln^2(1+x)}{x}\,dx = \frac{1}{4}\zeta(3).$$

Now we are all set to go in computing our alternating Euler sum!
    Using (4.4.16) in (4.4.21), we have

$$\frac{1}{4}\zeta(3) = \int_0^1 \frac{2\sum_{q=1}^{\infty}(-1)^{q-1}x^{q+1}\dfrac{h(q)}{q+1}}{x}\,dx = 2\sum_{q=1}^{\infty}(-1)^{q-1}\frac{h(q)}{q+1}\int_0^1 x^q\,dx$$

$$= 2\sum_{q=1}^{\infty}(-1)^{q-1}\frac{h(q)}{q+1}\left(\frac{x^{q+1}}{q+1}\right)\Big|_0^1$$

or

(4.4.22)
$$\frac{1}{4}\zeta(3) = 2\sum_{q=1}^{\infty}(-1)^{q-1}\frac{h(q)}{(q+1)^2}.$$

Next, change the index in (4.4.22) to $k = q + 1$ to get

$$\frac{1}{4}\zeta(3) = 2\sum_{k=2}^{\infty}(-1)^{k-2}\frac{h(k-1)}{k^2} = 2\sum_{k=2}^{\infty}(-1)^{k-2}\frac{h(k)-\dfrac{1}{k}}{k^2}$$

or

$$\frac{1}{4}\zeta(3) = 2\left\{\sum_{k=2}^{\infty}(-1)^{k-2}\frac{h(k)}{k^2} - \sum_{k=2}^{\infty}(-1)^{k-2}\frac{1}{k^3}\right\}.$$

If we start the index in both sums from $k = 1$, this becomes

$$\frac{1}{8}\zeta(3) = \left[\sum_{k=1}^{\infty}(-1)^{k-2}\frac{h(k)}{k^2} + h(1)\right] - \left[\sum_{k=1}^{\infty}(-1)^{k-2}\frac{1}{k^3} + 1\right],$$

which becomes, if we write q instead of k for the index (and because $h(1) = 1$ and $(-1)^{k-2} = -(-1)^{k-1}$),

$$\frac{1}{8}\zeta(3) = -\sum\nolimits_{q=1}^{\infty}(-1)^{q-1}\frac{h(q)}{q^2} + \sum\nolimits_{q=1}^{\infty}(-1)^{q-1}\frac{1}{q^3}.$$

Thus,

$$(4.4.23) \qquad \sum\nolimits_{q=1}^{\infty}(-1)^{q-1}\frac{h(q)}{q^2} = \sum\nolimits_{q=1}^{\infty}(-1)^{q-1}\frac{1}{q^3} - \frac{1}{8}\zeta(3).$$

We've seen the sum on the right-hand side of (4.4.23) before, as the alternating zeta function in (2.1.9) and (2.1.10), where we showed that

$$\sum\nolimits_{q=1}^{\infty}(-1)^{q+1}\frac{1}{q^s} = (1-2^{1-s})\zeta(s).$$

The alternating series in the sum of (4.4.23) is the $s = 3$ case (notice that $(-1)^{q+1} = (-1)^{q-1}$) and so

$$\sum\nolimits_{q=1}^{\infty}(-1)^{q-1}\frac{1}{q^3} = (1-2^{-2})\zeta(3) = \frac{3}{4}\zeta(3).$$

Using this in (4.4.23), we at last arrive at our answer:

$$(4.4.24) \qquad \sum\nolimits_{q=1}^{\infty}(-1)^{q-1}\frac{h(q)}{q^2} = \frac{3}{4}\zeta(3) - \frac{1}{8}\zeta(3) = \frac{5}{8}\zeta(3) = 0.75128\ldots.$$

We can check this theoretical result with a direct calculation of the alternating series, and this is done by the code **alt1775.m**, which is a simple variation of **euler1775.m**. As Figure 4.4.1 shows, using just the first 100 terms, the series converges fairly rapidly, with the value of the 100th partial sum being $0.75102\ldots$, in pretty good agreement with theory.

```
%alt1775.m
q=1;h(q)=1;f=-1;
for q=2:100
    h(q)=h(q-1)+1/q;
end
s(1)=1;
for q=2:100
    s(q)=s(q-1)+f*h(q)/q^2;
    f=-f;
end
q=[1:1:100];
plot(q,s,'-k')
xlabel('number of terms')
ylabel('sum')
s(100)
```

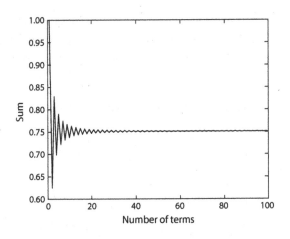

FIGURE 4.4.1.

The convergence of the sum in (4.4.24).

Well, okay, maybe this is enough on Euler sums. I think you can
now appreciate how these sorts of calculations really have no end.
There will never be a final Euler sum and so no end to good homework problems for professors to assign to students eager to tackle
ever more challenging calculations. There is, however, an end to this
book, and we have arrived at it. To celebrate that event, here's a final
challenge problem for you to try your hand at.[5]

The Final Challenge: In a paper[6] written by a father-son Canadian
team of mathematicians, we read that in 1993, one of their under-
graduate students (Enrico Au-Yeung, who is now (2021) a professor
of mathematics at DePaul University in Chicago) "conjectured on
the basis of a computation of 500,000 terms that $\sum_{q=1}^{\infty} \frac{h^2(q)}{q^2} = \frac{17}{4}\zeta(4)$."
The elementary (if lengthy) analysis I took you through a while ago
lay far in the future of 1993, and so the professors wrote that "our
first impulse was to perform a higher-order computation [out to 25
digits, compared to the student's mere 5 digits] to show [the claim]
to be false." To their surprise, however, the numbers checked, and
so "now armed with the *assurance*[7] [my emphasis] that the result was
true, we were prepared to look for [a formal derivation]." They were
able to do that, too (but it wasn't easy). However, they did include
the following provocative comment: "We did not know at the time
that P. J. De Doelder[8] had established [two years earlier, in 1991] that
$\sum_{q=1}^{\infty} \frac{h^2(q)}{(q+1)^2} = \frac{11}{4}\zeta(4)$ from which the [conjectured sum] is an immediate consequence." (See the following box for a comment on deriving
De Doelder's expression.) The professors didn't show how to do
that, however, and so there's your final challenge: Derive Au-Yeung's

5. You can find many more examples of Euler sum calculations, of much greater
complexity than I've shown you here, in the book by Cornel Ioan Vălean, *(Almost)
Impossible Integrals, Sums, and Series* (Springer, 2019).

6. David Borwein (born 1924) and Jonathan M. Borwein (1951–2016), "On an
Intriguing Integral and Some Series Related to ζ(4)," *Proceedings of the American
Mathematical Society*, April 1995, pp. 1191–1198.

7. And so here we see an example of how (some) mathematicians have moved
into the camp of physicists and engineers in believing computer-based numerical
checks, using large numbers of digits, may have real value in mathematical research.

8. Pieter J. De Doelder (1919–1994) was a Dutch mathematician.

conjectured identity directly from De Doelder's expression. Hint: This isn't nearly as long as the analysis we did in the text!

The same year the Borwein and Borwein paper cited in this chapter appeared, they (and a colleague) wrote another paper that, in addition to deriving De Doelder's 1991 expression, derived another expression much like De Doelder's that involved $\zeta(3)$:

$$\sum_{q=1}^{\infty} \frac{h^2(q)}{(q+1)^4} = \frac{2}{3}\zeta(6) - \frac{1}{3}\zeta(2)\zeta(4) + \frac{1}{3}\zeta^3(2) - \zeta^2(3)$$

and so

$$\zeta(3) = \sqrt{\frac{2}{3}\left(\frac{\pi^6}{945}\right) - \frac{1}{3}\left(\frac{\pi^2}{6}\right)\left(\frac{\pi^4}{90}\right) + \frac{1}{3}\left(\frac{\pi^2}{6}\right)^3 - \sum_{q=1}^{\infty}\frac{h^2(q)}{(q+1)^4}} =$$

$$\sqrt{\pi^6\left(\frac{2}{2,835} - \frac{1}{1,620} + \frac{1}{648}\right) - \sum_{q=1}^{\infty}\frac{h^2(q)}{(q+1)^4}}.$$

(You can find a derivation of this expression in Borwein, Borwein, and Girgensohn, "Explicit Evaluation of Euler Sums," *Proceedings of the Edinburgh Mathematical Society* 38, 1995, pp. 277–294.) Executing an altered (in the obvious way) version of the *MATLAB* code **numsquared.m**, using the first 10,000 terms of the sum, resulted in the estimate $\zeta(3) = 1.20205690317\ldots$, which has the first 10 decimal digits correct.

Epilogue

Every time I make a discovery I suddenly become very happy and that lasts for a long while. The philosophers looking for the key to happiness should begin to study mathematics, and probably they would be very surprised to see how well-connected happiness and mathematics are.

—From a December 2018 e-mail sent to the author by a mathematician in Romania[1]

How to end a book like this one, which deals with an unsolved problem that has stumped mathematicians for centuries? That question puzzled me from the very first day I started to write, because it seems unfair to leave you without at least a hint as to what I think the prospects for an eventual solution may be. Since the problem of $\zeta(3)$ defeated even the great Euler, you shouldn't be surprised to read that my first response is that it's going to take a mind that surpasses that of Euler's, a requirement not satisfied by even one of the numerous brilliant mathematicians who have pondered $\zeta(3)$ over the nearly two-and-a-half centuries since his death. So, given that, how much *more* powerful than Euler's brain will be the one to finally tell us what $\zeta(3)$ equals? If one is a pessimist, then the immediate answer is that it will take a "Chuck Norris" brain to solve for $\zeta(3)$, which is a pop-culture way of saying it will require a supernatural mentality (see the box).

1. Cornel Ioan Vălean, author of *(Almost) Impossible Integrals, Sums, and Series* (Springer, 2019).

ALERT: Everything in this box is just for fun!

Chuck Norris (born 1940) is an American martial arts action movie actor famous for being super-tough, and that reputation has spawned an amusing "literary" genre called *Chuck Norris math jokes*. Here are some of the funnier ones:

"Chuck Norris is so tough that even though $\sqrt{-1}$ is imaginary it's still afraid of Chuck."

"Chuck Norris is so fast that one day he counted by ones from one to infinity—twice."

"Chuck Norris can square the circle using only a pencil and his magnificent hand-eye coordination."

"Dividing by zero is easy for Chuck Norris, which is why he can simultaneously solve the equations of parallel lines."

"Chuck Norris doesn't differentiate because he's so tough he dis*integrates* everything."

"Chuck Norris knows which is bigger, pi or pi with its digits reversed, because he once calculated the exact value of each."

"Chuck Norris can draw a triangle with four sides—and *nobody* had better say he can't!"

Well, you get the drift. Surely, goes this theory, Chuck Norris would find the puzzle of $\zeta(3)$ to be mere child's play, and if we could only get him to stop wasting his time beating up evil movie thugs, and to turn his attention to $\zeta(3)$, then he surely would, overnight, become the world's most acclaimed mathematician. And, you have to admit, it would be great fun to hear scholarly mathematicians speak at math conferences in hushed, reverential voices on what would, without a doubt, become known as *Chuckie's number*.

If you're an optimist, however, you'll find some solace in some famous words by the late Hungarian-American, Harvard-based mathematician Raoul Bott, who was quoted at the beginning of this book. To paraphrase Bott, when he was once asked how great mathematics is done, he replied that there are two tried-and-true approaches. The first is the obvious one of just being smarter than everybody else. That, unfortunately, works only for a very small number of people (if you're literal minded, just one person can be smarter than everybody else). Of course, all it will take to solve for $\zeta(3)$ *is* one person, so perhaps that might work. Bott's other way to success is to be single-mindedly obsessed and simply plug away for however long it takes, even the rest of your life. After all, if Euler hadn't lost his eyesight and then simply run out of time, maybe he would have eventually found $\zeta(3)$. And don't forget, after Euler finally calculated the exact value for $\zeta(2)$, mathematicians started finding easier derivations, some of which are understandable by high school AP-calculus students. Maybe it will be the same with $\zeta(3)$. Maybe, in fact, somebody will find one of the "easy" derivations for $\zeta(3)$ right off the bat!

Bott didn't come straight out and say it, but one way a "persistent, high-energy" mind might look for $\zeta(3)$ is to simply guess it. I know that sounds pretty crude (because it is), but who cares? If by randomly combining lots of math's well-known constants (π, e, $\ln(2)$, $\sqrt{3}$, and so on), and using various functions (logarithmic, trigonometric, factorial, exponential, and so on), maybe by sheer luck you'll stumble onto a combination that endlessly churns out the correct digits of $1.2020569\ldots$.[2] You might feel better about this ad hoc approach by knowing that Euler wasn't so snobby as to be above using it himself. Recognizing that $\zeta(2n)$ has the form of $\frac{p}{q}\pi^{2n}$, where p and q are integers, Euler wondered if $\zeta(3)$ might equal $\frac{p}{q}\pi^3$ and so,

2. As I bring the writing of this book to an end, the value of $\zeta(3)$ has been computed out to at least 10^{12} (yes, a trillion) digits. To learn more about how that was done, type "y-cruncher" into Google. You'll be directed to numerous sites that will tell you all about the computer code called *y-cruncher*, which has also computed the digits of several other well-known constants (like Euler's gamma) out to an equally fantastic number of digits. Indeed, it was gamma, γ, which looks (sort of) like a *y*, that gave *y-cruncher* its name.

using the numerical value of $\zeta(3)$, calculated $\frac{q}{p} = \frac{\pi^3}{\zeta(3)}$ with the hope the result would offer a clue for what p and q might be.

This, in fact, didn't result in much of anything, but my point is that even the great Euler couldn't resist trying to reverse-engineer $\zeta(3)$. With the now easy availability of high-speed electronic computers, perhaps a systematic evaluation of millions upon millions of different combinations per second, using lots of math constants and functions, might lead to something. Or maybe not. I honestly don't give it much hope, but who really knows without trying? And, if the guessing approach *does* work for you, please remember (when you follow in Bott's footsteps and accept the Wolf Prize of 20??) where you got your inspiration!

The difficulty of the zeta-3 problem is the basis for this little amusement, one I would occasionally tell my own students: "Two cannibals are eating a badly cooked clown when one of them turns to the other, a frown on his face, and asks 'Does this taste just a little bit funny to you?'" I would set the stage for this admittedly tasteless joke by telling the class I had heard it from a mathematician who had been driven a bit looney after 50 years of failing to solve the zeta-3 problem. My students would either half-laugh or (more often) groan at that, but all agreed on one point—to repeat a joke that awful, one *had* to be just a bit unbalanced (a conclusion, I suspect, that was a message meant for me).

A modern unsolved problem that lends itself to computer study is the *Collatz conjecture*, made in 1937 by the German mathematician Lothar Collatz (1910–1990). While easily understandable by a grammar school student, it has stumped mathematicians ever since its appearance. Imagine a sequence of positive integers x_1, $x_2, x_3, \ldots x_k, x_{k+1}, \ldots$, where

$$x_{k+1} = \begin{cases} \dfrac{x_k}{2} & \text{if } x_k \text{ is even,} \\ 3x_k + 1 & \text{if } x_k \text{ is odd.} \end{cases}$$

The conjecture is that for x_1 equal to any positive integer greater than 1, the sequence will eventually generate 1 (and then loop endlessly through 4,2,1,4,2,1, . . .). Figure E.1 shows a plot of the sequence for x_1 = 97. There is an enormous body of computer studies that all support the conjecture but, as with the Riemann hypothesis, those vast numerical results *prove* nothing.

In any case, to help you avoid the fate of my mathematician friend as *you* slog away on $\zeta(3)$, some words from the French mathematician Georges-Louis Leclerc (1707–1788), Comte de Buffon, might be good to keep in mind: "Never think that God's delays are God's denials. Hold on; hold fast; hold out. Patience is genius!" In other words, just because God made $\zeta(3)$ a very hard problem doesn't mean He necessarily made it an impossible one. In particular, the failure of Euler to compute $\zeta(3)$ doesn't mean all hope is lost, as Euler was not infallible.

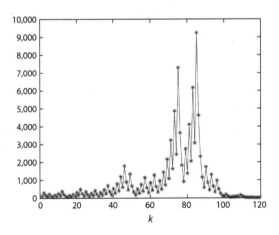

FIGURE E.1.

The Collatz sequence with starting value x_1 = 97.

As an example of that, in 1769 Euler observed $x^2 + y^2 = w^2$ has integer solutions ($3^2 + 4^2 = 5^2$), while $x^3 + y^3 = w^3$ does not (think of Fermat's Last Theorem), and yet $x^3 + y^3 + z^3 = w^3$ does ($3^3 + 4^3 + 5^3 = 6^3$). On this rather thin evidence, Euler then conjectured that at least n nth powers are required to form a sum equal to an nth power. So, to have an all-integer solution for the $n = 5$ case, for example, Euler thought one needed to have at least five integers on the left, that is, $x^5 + y^5 + z^5 + u^5 + v^5 = w^5$. In 1966, however, a computer search showed Euler's conjecture to be false, with the discovery of the four-integer counterexample $27^5 + 84^5 + 110^5 + 135^5 = 144^5$. Euler was fleet of foot, yes, but every now and then he could still stub a toe.[3] Did he somehow miss a crucial observation in his long hunt for $\zeta(3)$?

So, here's my parting word, particularly to all the younger readers of this book, who, while perhaps not smarter than everybody else, are persistent. *Start plugging*! Your obvious reward will be fame, of course, but the even greater reward will be what the opening quotation from my Romanian correspondent proclaimed. The Hungarian-born American mathematician Paul Halmos (1916–2006) put the situation this way, in a 1990 interview: "What's the best part of being a mathematician? I'm not a religious man, but it's almost like being in touch with God when you're thinking about mathematics. God is keeping secrets from us [think $\zeta(3)$], and it's fun to try to learn some of the secrets."[4]

---

3. The year after he made his conjecture, Euler offered the first proof that there are no integer solutions to $x^3 + y^3 = w^3$. (Fermat, himself, had already shown the same for $x^4 + y^4 = w^4$.) Alas, there were some subtle difficulties with Euler's proof that were later cleaned up by others. Euler's basic idea was okay, but there were some missteps.

4. Don Albers, "In Touch with God: An Interview with Paul Halmos," *College Mathematics Journal*, January 2004, pp. 2–14.

Appendix 1

Solving the Impossible by Changing the Rules

> "This is beautiful," Rebus said to himself. He hadn't just squared
> the circle, he'd created an unholy triangle out of it.
>
> —Inspector John Rebus suddenly unravels the puzzles of a multiple
> murder investigation in Ian Rankin's 1994 novel *Mortal Causes*

In this appendix I'll elaborate, just a bit, on the nature of unambiguous mathematical questions that, despite being crystal clear in what they ask for, nevertheless have no solutions. I first mentioned this issue at the end of Section 1.2, concerning the $\zeta(3)$ problem, with an intentionally cryptic remark about the impossibility of solving certain ancient geometric construction problems (trisecting an angle, squaring a circle, and doubling a cube), problems dating from centuries before Christ.

On the opening pages of his excellent book[1] on these problems, the American mathematician Nicholas Kazarinoff (1921–1991) included this insightful passage:

---

1. Nicholas Kazarinoff, *Ruler and the Round: Classic Problems in Geometric Constructions* (Dover, 2003), first published in 1970. That book is written at the same level as this one (high school geometry and AP-calculus). You can find more discussions on the history of these ancient construction problems in David S. Richeson, *Tales of Impossibility: The 2000-Year Quest to Solve the Mathematical Problems of Antiquity* (Princeton University Press, 2019).

One must not confuse the impossibility of a geometric construction with an unsolved problem—or with the insolvability of a problem! Consider the following example: ... construct a square whose side length is a whole number of units and whose area is two square units. Clearly, there exists no such square.... On the other hand, if we change our rules slightly and admit as candidates for solutions to our problem squares of any side length, then we can solve the problem affirmatively. (Given a straight line segment *AB* of unit length, we construct a second segment *AC* perpendicular to *AB* at *A* and also of unit length. Then *BC* is a side of a square of area 2. The length of *BC* is $\sqrt{2}$, which is not a whole number.)

To follow up on Kazarinoff's words, consider the problem of squaring a circle (using straightedge and compass alone, construct a square with area equal to that of any given circle). This particular problem is so famous that it has entered into general use, even among non-mathematicians, as *the* metaphor (as in the quotation that opens this appendix) for achieving the impossible. To be quite specific, suppose the given circle has unit radius and so its area is π. The specific problem, then, is to construct a square with side length $\sqrt{\pi}$. Nobody, alas, could find a way to do that, even after thousands of years of trying—and then, in 1882, the German mathematician Ferdinand Lindemann (1852–1939) finally discovered the reason behind that colossal failure: π is what is called a *non-constructable number*, which means that, given a line segment of unit length, it is impossible to construct (with just a straightedge and compass) a line segment of length π (you can find more on Lindemann's proof in Kazarinoff's book; see the first note in this appendix). This instantly showed the impossibility of squaring a circle. Here's why.

Suppose we *have* succeeded, somehow, in constructing a length $\sqrt{\pi}$. Then, using a simple, high school construction that generates a length that is the square of any given length,[2] we could construct

2. Do you see how to construct a length x^2 from a given length x (also given is the unit length, which simply sets the scale of the construction)? This question has nothing to do with $\zeta(3)$—at least, I don't think it does—but it would be cruel to leave you hanging, and so you'll find a solution in the box at the end of this appendix.

the length π. But Lindemann proved that is simply not possible. So the initial supposition of having earlier constructed the length $\sqrt{\pi}$ must be in error. That's it!

The impossibility of squaring a circle is an actual impossibility, however, only if we are constrained to using the traditional construction tools of straightedge and compass. If we change the rules just a bit, then we can square a circle. The use of a straightedge means we can draw "curves" called straight lines, and the use of a compass means we can draw curves called circles. Let's now suppose we have, in addition, a third instrument that draws the *quadratrix* curve, a mathematical creation dating from circa 450 B.C. To go into the details of the quadratrix, and the instrument that draws it (it is not difficult to make), would take us too far afield from the theme of this book (see Kazarinoff's book, pp. 28 and 60–61, cited in the first note of this appendix), but with it squaring a circle is now possible.

Well, does that hold out the tantalizing possibility that the $\zeta(3)$ problem could also be made to have a solution if we could, in some sense, somehow change the rules? When I wrote the first draft of this appendix, my answer was no, because for the square-a-circle problem, the rules were in the form of specifying the drawing instruments we could use. Changing the rules in that case simply meant changing the allowed drawing instruments, and that is obviously physically possible to do. My initial thoughts on the same issue for the $\zeta(3)$ problem, however, were that the rules we would then be playing by are the very laws of mathematics, and how can we possibly change them? Like the proclamation in Ralph Waldo Emerson's poem "The Past," on the rigidity of history ("All is now secure and fast/Not the gods can shake the Past"), the rules of mathematics seem to be equally unalterable.

But after getting back the comments from the initial reviews of my proposal for this book, one of those comments in particular caught my eye. A reviewer (who, alas, elected to remain anonymous) wrote that perhaps the difficulty in expressing $\zeta(3)$ as a combination of certain numbers (the integers, π, e, and so on) is because our list of candidate numbers is missing one (or perhaps even more) possibilities. That is, numbers that we simply haven't yet discovered; for example, the zeros of some new function, like π is a zero of $\sin(x)$. That seemed to me to be something worth thinking about, as it holds out the hope that the solution to the puzzle of $\zeta(3)$ may suddenly appear one day from out of a seemingly distant area of mathematics.[3]

And doesn't that strike you as something a tough and demanding (but ultimately non-malicious) Creator would do? Let's keep our fingers crossed!

In Figure A1.1 you see the right triangle **AOB**, where the length of **OA** is the given unit length, and the length of **OB** (constructed perpendicular to **OA** at **O**) is the given x. In this construction we'll assume that $x > 1$ (which is the case for $x = \sqrt{\pi}$), but the case of

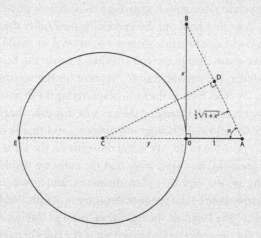

FIGURE A1.1.

How to construct x^2 from $x > 1$.

3. The idea of discovering a new number has long fascinated me, and long ago I wrote a short science fiction story with that as the "catch." Titled "Mathematical R&D," it originally appeared in the *IEEE Transactions on Aerospace and Electronic Systems*, January 1979, pp. 179–180, and is reprinted in my book, *Number-Crunching* (Princeton University Press, 2011), pp. 302–304.

$x < 1$ requires only a nearly trivial modification that I'll leave for you to think about. (The case of $x = 1$ is particularly trivial, since then $x^2 = x$, and you are done before you start.) The length of the hypotenuse **AB** is (from Pythagoras) $\sqrt{1+x^2}$. Now, bisect the hypotenuse (creating **D**), and extend the perpendicular bisector to intersect the extension of **OA** at **C**; denote the length of **OC** by y.

Denote the angle α (common to the right triangles **AOB** and **ACD**) as shown in the figure. Clearly, $\cos(\alpha) = \dfrac{1}{\sqrt{1+x^2}} = \dfrac{\frac{1}{2}\sqrt{1+x^2}}{1+y} = \dfrac{\sqrt{1+x^2}}{2(1+y)}$. Thus, $2(1 + y) = 1 + x^2 = 2 + 2y$, or $x^2 = 1 + 2y$. So, using **C** as the center of a circle with radius y, draw that circle and call its intersection with the extension of **OA** point **E**. The length of **ECOA** is $1 + 2y$ and so is the desired x^2. (Now, do you see how this construction changes for the $x < 1$ case?)

One final comment on impossibility in mathematics. Are we, today, living too late in the history of mathematics to enjoy a contemporaneous problem of impossibility? That is, is there nothing left for a modern mathematician to do in this area? The answer is a resounding *NO*, and I describe some examples of that in Appendix 5, from the mathematical theory of computer science.

Appendix 2

Evaluating $\int_0^\infty e^{-t^2}\,dt$ and $\int_0^\infty e^{-pt^2-\frac{q}{t^2}}\,dt$

The value I of an integral certainly doesn't depend on the shape of the particular squiggle of ink we use to denote the dummy variable of integration. That is, using t or x or y doesn't matter:

$$I = \int_0^\infty e^{-t^2}\,dt = \int_0^\infty e^{-x^2}\,dx = \int_0^\infty e^{-y^2}\,dy.$$

Now, consider the double integral

$$\int_0^\infty e^{-x^2}\left\{\int_0^\infty e^{-y^2}\,dy\right\}dx = \int_0^\infty e^{-x^2}\,dx\int_0^\infty e^{-y^2}\,dy = I^2$$

where we can justify moving the inner y-integral out of the double integral because that inner integrand has no x dependence. But there is nothing that says we couldn't instead move the inner y-integral *into* the x-integral just as well. That is, we can also write I^2 as

$$\int_0^\infty e^{-x^2}\left\{\int_0^\infty e^{-y^2}\,dy\right\}dx = \int_0^\infty\int_0^\infty e^{-(x^2+y^2)}\,dxdy.$$

That is,

$$I^2 = \int_0^\infty\int_0^\infty e^{-(x^2+y^2)}\,dxdy.$$

We physically interpret the I^2 integral as follows. The double integral's region of integration is the entire first quadrant of the xy-plane, with the physical significance of $dxdy$ being the differential area patch in Cartesian coordinates; see note 10 in Chapter 2. Now, the numerical value of the double integral, I^2, certainly doesn't depend on the particular coordinate system we happen to use. If we move to polar coordinates (r, θ), we have $x = r\cos(\theta)$ and $y = r\sin(\theta)$ and so $x^2 + y^2 = r^2$, the differential area patch in polar coordinates, is given by $rdrd\theta$, and to integrate over the entire first quadrant, we use $0 \leq r \leq \infty$ and $0 \leq \theta \leq \frac{\pi}{2}$. Since in either coordinate system, the double integral must come out to the same numerical value, we have

$$I^2 = \int_0^\infty \int_0^{\pi/2} e^{-r^2} r dr d\theta = \int_0^\infty e^{-r^2} r \left\{ \int_0^{\pi/2} d\theta \right\} dr$$

$$= \frac{\pi}{2} \int_0^\infty e^{-r^2} r dr = \frac{\pi}{2} \left[\left(-\frac{1}{2} e^{-r^2} \right) \Big|_0^\infty \right] = \frac{\pi}{4}$$

which says

$$I = \int_0^\infty e^{-t^2} dt = \frac{1}{2}\sqrt{\pi},$$

a result used in the derivation of (1.4.6). Every engineer and physicist, and certainly all mathematicians, should know this gem of analysis, the evaluation of an integral that occurs in countless scientific applications and theoretical situations. The calculation of this integral is occasionally attributed to Gauss (see note 1 in Chapter 1), but Gauss himself always credited the French mathematician Pierre-Simon Laplace (1749–1827) who, in fact, did it in 1774 during his early work in probability theory. It is, in fact, commonly called the *probability integral*. For an entirely different (less physical, more mathematical) derivation, see my book, *Hot Molecules, Cold Electrons* (Princeton University Press, 2020), pp. 185–187.

Now that we have a formula for the probability integral, we can extend it to the more general

$$I = \int_0^\infty e^{-pt^2 - \frac{q}{t^2}} dt,$$

which reduces to the probability integral as the special case of $p = 1$ and $q = 0$. This generalized result, you'll recall, played a central role in doing the Feynman-Hibbs integral in the Preface. Here's how to do the generalization with just AP-calculus.

To start, let me remind you of note 24 in Chapter 1, and of the box that ends Section 1.6. There I showed you a simple form of Feynman's trick of differentiating an integral with respect to a parameter in the integrand. We can use that idea here to do our new integral. We begin by changing variable to $u = t\sqrt{p}$ ($dt = \frac{du}{\sqrt{p}}$) and so

$$I = \int_0^\infty e^{-u^2 - \frac{qp}{u^2}} \frac{du}{\sqrt{p}} = \frac{1}{\sqrt{p}} \int_0^\infty e^{-u^2 - \frac{qp}{u^2}} du = \frac{1}{\sqrt{p}} I_2$$

where

$$I_2 = \int_0^\infty e^{-u^2 - \frac{qp}{u^2}} du.$$

If we write $k = \sqrt{qp}$, then we have

$$I_2(k) = \int_0^\infty e^{-u^2 - \frac{k^2}{u^2}} du$$

and so, if we differentiate with respect to k and assume that we can reverse the order of differentiation and integration, we have

$$\frac{dI_2(k)}{dk} = \int_0^\infty -\frac{2k}{u^2} e^{-u^2 - \frac{k^2}{u^2}} du.$$

If we next change variable to $y = \frac{k}{u}$ (and so $\frac{dy}{du} = -\frac{k}{u^2}$, or $du = -\frac{u^2}{k} dy$), then

$$\frac{dI_2}{dk} = -2k\int_\infty^0 \frac{1}{u^2} e^{\frac{k^2}{y^2}-\frac{k^2}{k^2/y^2}}\left(-\frac{u^2}{k}dy\right)$$

or

$$\frac{dI_2}{dk} = -2\int_0^\infty e^{-\frac{k^2}{y^2}-y^2}\,dy = -2I_2.$$

That is,

$$\frac{dI_2}{I_2} = -2dk$$

or, with C some constant, we have (after indefinite integration)

$$\ln(I_2) = -2k + \ln(C)$$

or

$$I_2(k) = Ce^{-2k}.$$

We can determine C by noticing that $I_2(0)$ is the probability integral. That is,

$$I_2(0) = C = \frac{1}{2}\sqrt{\pi}$$

and we thus have

$$I_2(k) = \frac{1}{2}\sqrt{\pi}e^{-2k} = \frac{1}{2}\sqrt{\pi}e^{-2\sqrt{qp}}.$$

Since

$$I = \frac{1}{\sqrt{p}} I_2$$

we immediately have our answer:

$$I = \int_0^\infty e^{-pt^2 - \frac{q}{t^2}} dt = \frac{1}{2} \sqrt{\frac{\pi}{p}} e^{-2\sqrt{qp}}.$$

As a quick check, if $q = p = 1$, our formula says $\int_0^\infty e^{-t^2 - \frac{1}{t^2}} dt = \frac{1}{2} \sqrt{\pi} e^{-2}$ $= \frac{\sqrt{\pi}}{2e^2} = 0.11993777196806\ldots$, while *MATLAB* says *integral(@(x) exp(-x.^2-1./(x.^2)),0,inf) = 0.11993777196806. . .*, which is pretty good agreement. If you are willing to be fearless, and to let p and/or q be imaginary constants, then some really spectacular results can be derived from our formula.

For example, as early as 1743 Euler became interested in the two definite integrals, $\int_0^\infty \cos(t^2) dt$ and $\int_0^\infty \sin(t^2) dt$, in connection with the physics of a coiled spring. After decades of effort, he finally (1781) evaluated both using his gamma function.[1] Ironically, despite that success these two integrals are, today, called the Fresnel integrals, after the French scientist Augustin Jean Fresnel (1788–1827), who encountered them in an 1818 study of the illumination intensity of optical diffraction patterns. (Note that Euler had evaluated the Fresnel integrals years before Fresnel was born!)

To simultaneously evaluate these two integrals with our formula, let $q = 0$ and $p = i = \sqrt{-1}$. Then our formula becomes, using (appropriately enough) Euler's identity,

$$\int_0^\infty e^{-it^2} dt = \int_0^\infty \cos(t^2) dt - i \int_0^\infty \sin(t^2) dt = \frac{1}{2} \sqrt{\frac{\pi}{i}}.$$

---

1. The details of how Euler used the gamma function to do these two integrals are discussed in my *Inside Interesting Integrals*, 2nd edition (Springer, 2020), p. 348.

But $i = e^{i\frac{\pi}{2}}$ and so

$$\int_0^\infty \cos(t^2)dt - i\int_0^\infty \sin(t^2)dt = \frac{1}{2}\sqrt{\frac{\pi}{e^{i\frac{\pi}{2}}}} = \frac{1}{2}\frac{\sqrt{\pi}}{e^{i\frac{\pi}{4}}} = \frac{\sqrt{\pi}}{2}e^{-i\frac{\pi}{4}}$$

$$= \frac{\sqrt{\pi}}{2}\left\{\cos\left(\frac{\pi}{4}\right) - i\sin\left(\frac{\pi}{4}\right)\right\} = \frac{\sqrt{\pi}}{2}\left\{\frac{1}{\sqrt{2}} - i\frac{1}{\sqrt{2}}\right\} = \frac{1}{2}\sqrt{\frac{\pi}{2}} - i\frac{1}{2}\sqrt{\frac{\pi}{2}}.$$

Equating real and imaginary parts, we then immediately have, just like that,

$$\int_0^\infty \cos(t^2)dt = \int_0^\infty \sin(t^2)dt = \frac{1}{2}\sqrt{\frac{\pi}{2}}.$$

I won't go through the details here,[2] but if you're looking for other similar, equally amazing exercises, first set $p = i$ and $q = -i$ in our formula and see if you can then show that

$$\int_0^\infty \cos\left(t^2 - \frac{1}{t^2}\right)dt = \int_0^\infty \sin\left(t^2 - \frac{1}{t^2}\right)dt = \frac{1}{2e^2}\sqrt{\frac{\pi}{2}}.$$

If you can do that, then set $p = q = i$ and see if you can show that

$$\int_0^\infty \cos\left(t^2 + \frac{1}{t^2}\right)dt = \sqrt{\frac{\pi}{2}}\cos\left(2 + \frac{\pi}{4}\right) \text{ and } \int_0^\infty \sin\left(t^2 + \frac{1}{t^2}\right)dt = \sqrt{\frac{\pi}{2}}\sin\left(2 + \frac{\pi}{4}\right).$$

If using imaginary values for p and q leaves you just a bit nervous, you are not alone. As an historian of mathematics recently wrote,[3]

A curious feature of mathematical analysis in the years around 1800 was the use of complex variables to evaluate real definite integrals. The practice had

2. You can find them in my *Inside Interesting Integrals*, pp. 349–350, 487.
3. Jeremy Gray, *The Real and the Complex: A History of Analysis in the Nineteenth Century* (Springer, 2015), pp. 59–60.

begun with Euler. . . . In his *Mémoire* on this topic that he presented in 1814 [the French mathematician Augustin-Louis Cauchy (1789–1857)] commented that many of the integrals had been evaluated for the first time "by means of a kind of induction" based on "the passage from the real to the imaginary" and that no less a figure than Laplace had remarked that the method "however carefully employed, leaves something to be desired in the proofs of the results." Cauchy accordingly set himself the task of finding a "direct and rigorous analysis" of this dubious passage.

Cauchy's 1814 *Mémoire* was the prelude to his later (1825) magnificent development of contour integration in the complex plane, about which you can read (in some detail) in my *Inside Interesting Integrals*, pp. 351–422. In advanced work on the mysteries of $\zeta(3)$, contour integrals abound, but such sophisticated doings are for another book.

Appendix 3

Proof That $\sum_{q=1}^{\infty} \sum_{\substack{n=1 \\ n \neq q}}^{\infty} \frac{1}{qn(n-q)}$ Equals Zero

To establish the claim, we start with Figure A3.1, which shows the points with positive integer coordinates in the first quadrant of the qn-plane. Each such point is associated with a term in the double sum, with the exception of those points on the diagonal line $n = q$ (because of the $n \neq q$ condition). The shaded triangular region labeled $R_{n>q}$ is the set of all points associated with the positive terms in the double sum, and the shaded triangular region labeled $R_{n<q}$ is the set of all points associated with the negative terms in the double sum. It is clear, by both the symmetry of the two regions and by the general form of the terms, that the sums over $R_{n>q}$ and $R_{n<q}$ are negatives of each other. Note, however, that at this point we cannot argue that these two individual sums add together to give zero. That's because if each individual sum is infinite, then their sum is $\infty - \infty$, which is indeterminate. We can conclude that the sum of our two individual sums is zero only if the individual sums are negatives and equal in *finite* magnitude (our assumption, you'll recall, in the discussion in Section 4.2).

It will suffice to show that the sum over $R_{n>q}$ is finite. Since the terms of that double sum are all positive and continually decrease toward zero as n and q increase, we can use the so-called *integral test* from calculus, which says that if we treat q and n as continuous

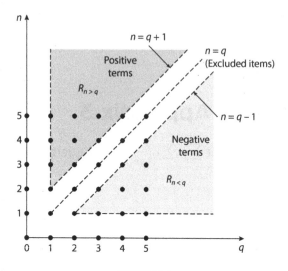

FIGURE A3.1.
The geometry of the double sum $\sum_{q=1}^{\infty}\sum_{n=q}^{\infty}\frac{1}{qn(n-q)}$.

variables, then $\iint_{R_{n>q}}\frac{1}{qn(n-q)}dqdn<\infty$ means that the double sum is also finite. So, that's our problem here, the evaluation of the double integral

$$\iint_{R_{n>q}}\frac{1}{qn(n-q)}dqdn=\int_1^{\infty}\frac{1}{q}\left\{\int_{q+1}^{\infty}\frac{1}{n(n-q)}dn\right\}dq$$

where the limits on the q and n integrations are obvious from Figure A3.1. Since

$$\frac{1}{n(n-q)}=\frac{1}{q}\left\{\frac{1}{n-q}-\frac{1}{n}\right\}$$

then

$$\iint_{R_{n>q}}\frac{1}{qn(n-q)}dqdn=\int_1^{\infty}\frac{1}{q^2}\left\{\int_{q+1}^{\infty}\left\{\frac{1}{n-q}-\frac{1}{n}\right\}dn\right\}dq.$$

Now,

$$\int_{q+1}^{\infty}\left\{\frac{1}{n-q}-\frac{1}{n}\right\}dn = \int_{q+1}^{\infty}\frac{dn}{n-q}-\int_{q+1}^{\infty}\frac{dn}{n}.$$

For the first integral on the right, change variable to $u = n - q$ (and so $du = dn$). Then

$$\int_{q+1}^{\infty}\left\{\frac{1}{n-q}-\frac{1}{n}\right\}dn = \int_{1}^{\infty}\frac{du}{u}-\int_{q+1}^{\infty}\frac{dn}{n}=\int_{1}^{q+1}\frac{dn}{n}$$

$$= \ln(n)\big|_{1}^{q+1}=\ln(q+1).$$

Thus,

$$\iint_{R_{n>q}}\frac{1}{qn(n-q)}dqdn = \int_{1}^{\infty}\frac{\ln(q+1)}{q^2}dq \le \int_{1}^{\infty}\frac{\ln(2q)}{q^2}dq$$

where the last integral and inequality follow because, over the entire interval of integration, it is clearly true that $q + 1 \le 2q$. So

$$\iint_{R_{n>q}}\frac{1}{qn(n-q)}dqdn \le \int_{1}^{\infty}\frac{\ln(2)+\ln(q)}{q^2}dq$$

$$= \ln(2)\int_{1}^{\infty}\frac{dq}{q^2}+\int_{1}^{\infty}\frac{\ln(q)}{q^2}dq.$$

Since $\int_{1}^{\infty}\frac{dq}{q^2}=\left\{-\frac{1}{q}\right\}\big|_{1}^{\infty}=1$, and $\int_{1}^{\infty}\frac{\ln(q)}{q^2}dq=\left\{-\frac{\ln(q)}{q}-\frac{1}{q}\right\}\big|_{1}^{\infty}=1$, then

$$\iint_{R_{n>q}}\frac{1}{qn(n-q)}dqdn \le \ln(2)+1 < \infty$$

and we are done.

Appendix 4

Double Integration Reversal Isn't Always Legal

We know that the evaluation or even only the reduction of multiple
integrals generally presents very considerable difficulties.

—Gustav Dirichlet (1839)

By the time you've gotten to the mid-point of this book, you will
have encountered numerous occasions where, in a double integral,
the order of the two integrations has been reversed. In none of those
occasions was any justification for the reversal provided. Instead, I
simply argued that such reversals are usually okay, and so we just
went ahead and did it to see what we got. Mathematicians are generally not amused by that, however, and so here I'll try to partially
atone for my sins of omission by admitting that

$$\int_a^b \int_c^d f(x,y)\,dxdy = \int_a^b \left\{ \int_c^d f(x,y)dx \right\} dy$$

and

$$\int_c^d \int_a^b f(x,y)\,dydx = \int_c^d \left\{ \int_a^b f(x,y)dy \right\} dx$$

may not be equal, even though we appear to be integrating the same function over the same finite rectangular region.

To demonstrate this, consider the classic example of $f(x,y) = \frac{x-y}{(x+y)^3}$ integrated over the unit square. Invoking my usual "let's just do it" argument, we'd write

$$\int_0^1 \left\{ \int_0^1 \frac{x-y}{(x+y)^3} dx \right\} dy = \int_0^1 \left\{ \int_0^1 \frac{x-y}{(x+y)^3} dy \right\} dx.$$

But au contraire! If we calculate the specific value of each side of this supposed equality, we'd find (perhaps to our amazement) that while the two sides do indeed have definite values, those two values are *not* equal. Here are the details of such calculations.

For $\int_0^1 \{\int_0^1 \frac{x-y}{(x+y)^3} dx\} dy$, let $t = x + y$ in the inner integral (where x is varying and y is held constant). Then $dx = dt$, and $\int_0^1 \frac{x-y}{(x+y)^3} dx = \int_y^{1+y} \frac{t-2y}{t^3} dt$ $= \int_y^{1+y} \frac{dt}{t^2} - 2y \int_y^{1+y} \frac{dt}{t^3} = (-\frac{1}{t})|_y^{1+y} - 2y(-\frac{1}{2t^2})|_y^{1+y} = (\frac{1}{y} - \frac{1}{1+y}) + y[\frac{1}{(1+y)^2} - \frac{1}{y^2}]$ or, after a little simple algebra, this reduces to $-\frac{1}{(1+y)^2}$. So $\int_0^1 \{\int_0^1 \frac{x-y}{(x+y)^3} dx\} dy = -\int_0^1 \frac{dy}{(1+y)^2}$. Let $t = 1 + y$, and this integral becomes $-\int_1^2 \frac{dt}{t^2} = -(-\frac{1}{t})|_1^2 = -(-\frac{1}{2}+1) = -\frac{1}{2}$. That is, $\int_0^1 \{\int_0^1 \frac{x-y}{(x+y)^3} dx\} dy = -\frac{1}{2}$. If you repeat this business[1] for $\int_0^1 \{\int_0^1 \frac{x-y}{(x+y)^3} dy\} dx$, you'll get $+\frac{1}{2}$. Do you think the reason for this lack of equality might be connected to the fact that the integrand blows up as we approach the lower left corner ($x = y = 0$) of the region of integration?

Maybe, but staying away from $x = y = 0$ doesn't necessarily avoid the problem. That's because if we change the region of integration to the infinite region $1 \leq x < \infty$, $1 \leq y < \infty$, we'll still experience a failure of equality. That is,

$$\int_1^\infty \left\{ \int_1^\infty \frac{x-y}{(x+y)^3} dx \right\} dy \neq \int_1^\infty \left\{ \int_1^\infty \frac{x-y}{(x+y)^3} dy \right\} dx$$

even though the integrand is now well behaved (that is, is continuous and finite) for all x and y. (Notice that the integrand, for both

---

1. Even easier is simply to notice that $\int_0^1 \{\int_0^1 \frac{x-y}{(x+y)^3} dy\} dx = -\int_0^1 \{\int_0^1 \frac{y-x}{(x+y)^3} dy\} dx$ which, by inspection, is the negative of the double integral we just did.

regions of integration, changes sign as we move about in each region—could that, perhaps, have something to do with the inequality of the integrals?) I'll let you fill in the details of showing the inequality (just mimic what we did for the finite region). These two examples have not rigorously identified the underlying requirements for a reversal of integration order to be valid, but instead only demonstrate that finding such requirements is an important (nontrivial) task, a task I'll leave for more advanced discussions in math books at a deeper level than is this one. Generally, however, if an integrand $f(x, y)$ is an *everywhere continuous, finite* function in a *finite* region of integration, reversal of the order of integration will not get you into trouble.

Appendix 5

Impossibility Results from Computer Science

This appendix is a continuation of the theme of Appendix 1, on mathematical problems that, despite clearly stating a well-formed question, have no solution. In Appendix 1 we discussed one such problem, an ancient problem from geometric construction, and here I'll show you additional similar problems from the modern discipline of computer science.

Modern electronic digital computers are of such massive capability that it is easy to suppose that there is no number-crunching problem that such machines couldn't, if given sufficient time, grind their way through. Unfortunately, that is just not true, and it has been known to be false since the English mathematician Alan Turing (1912–1954) proved it in 1936.[1] Turing arrived at his astonishing conclusion by the direct route of describing, in great detail, quite specific, particular computational problems that he showed are inherently unsolvable. His analyses hold true no matter how large the memory storage, or how fast the clock speed, of a computer may

---

1. Turing was a towering figure in the early days of computer science, and he played a pivotal role in the breaking of the German Enigma code of World War II. He was also fascinated by the Riemann hypothesis (concerning where in the complex plane the zeros of the zeta function are located), and you can find more on that in Andrew Hodges, *Alan Turing: The Enigma* (Simon and Schuster, 1983); and David Leavitt, *The Man Who Knew Too Much: Alan Turing and the Invention of the Computer* (W. W. Norton, 2006).

be. Make both those numbers a thousand million billion trillion times greater than they are for the most powerful computer in existence today, and Turing's problems continue to remain unsolvable. Even the eventual development of a quantum computer will not change this claim. I'll start with discussions of two of Turing's problems and end with some commentary on the question of the solvability (or not) of the Riemann hypothesis concerning the location of the complex zeros of the zeta function.

As the first example of what Turing did, consider this claim: There are an infinity of real numbers that are impossible to compute. On the face of it, this seems to be an outrageous statement, but it is, in fact, not at all difficult to prove. Here's how.

To start, let me make a few preliminary observations about the concept of infinity. We all know it's "big," but that doesn't even begin to get at the mathematics of what it means to say something is *infinite* in size. Most people, when asked to give an example of an infinite number of things, will probably reply "All the integers." That's correct, too, as the integers form what mathematicians call a *countably infinite set*. That's because we can literally count the integers, one, two, three, four, . . . , up to a billion, a trillion, and on and on and on. If you count off one integer each second, then I can tell you precisely the instant in the future you'll have counted up to any particular integer. The integers are infinite in number, yes, but they can be counted.

This is probably not so surprising to you, but there are other countable infinities that are surprising. For example, all the rational numbers (to be specific, the numbers from zero to one that are the ratios of integers) are a countable infinity. Why is that a surprising statement? It's a surprise, because unlike the integers, the rational numbers are *dense*. What a mathematician means by "dense" is that if you take any two rational numbers, no matter how close together they may be, there is another rational number between them (their average value). And so on, forever. There is no minimum separation of rational numbers; no matter how small a non-zero separation you name, there are two rational numbers that are even closer. In contrast, the integers are *not* dense, because there is a minimum separation between consecutive integers—dare I say it?—of one.

Nevertheless, despite their denseness, the rational numbers still form a countable infinity. This astonishing result, totally at odds with intuition (often called "common sense") was discovered by Cantor in 1874. Many mathematicians of his day thought Cantor was crazy, but it was they (not him) who were wrong (although, ironically, Cantor did die in a mental institution).[2] Cantor went on to show that the infinite set of all real numbers from zero to one does not form a *countable* infinity (for a proof of this, see my book, *The Logician and the Engineer*). Since there are just two categories of real numbers— the rational numbers (which form a countable infinity) and the irrational numbers (like π and $\sqrt{2}$)—then we immediately know that the irrational numbers must be so numerous as to form an uncountable infinity. With this conclusion, we can now prove the existence of Turing's non-computable numbers.

We start by visualizing a computer (imagine it to be as massively powerful as you like) being programmed to compute numbers using various algorithmic procedures whose details are unimportant. The programming is done in any language you wish (actually, to understand this argument, you really don't even have to know a programming language), with the only requirement being that the language uses a finite set of distinct symbols, for example, the 26 letters of the English alphabet, the 10 digits from 0 to 9, and a few additional special symbols like >, <, =, ^, (,). We suppose each new program we write with these symbols computes a new number. Let's now list all of these programs by symbol length.

That is, the first program on our list will be one symbol long. Since there is more than one program of length one, we'll list them in alphabetical order. Then we'll do the same for all programs of length two symbols, then three symbols, and so on. The number of programs of a given length is finite, since we have a finite set of

2. For Cantor's high school–level proof of the countable infinity of the rational numbers, see my book, *The Logician and the Engineer* (Princeton University Press, 2013), pp. 169–170.

distinct symbols at our disposal. It should be clear that the resulting infinitely long list of programs forms a countable infinity. But there is an uncountable infinity of real numbers, and so there must be an uncountable infinity of numbers left uncomputed, numbers we can't compute, because there simply aren't enough programs! It is important to understand that this result does not preclude being able to compute all the numbers that have a finite number of digits, like the first billion digits of $\zeta(3)$. There are a finite number of billion-digit numbers from one to two, for example, and since a countable infinity of programs is available, there would be no shortage of computational power. Turing's result comes into play only when we talk of computing all the exact values of the uncountable infinity of the real numbers.

Now, all this might strike you as being pretty academic and hugely abstract, as something far removed from the practical concerns of programmers and the nitty-gritty real-world of computer science. So, how about this "real-world" problem that haunts the nightmares of every computer coder: writing a program that contains the dreaded flaw of accidentally plunging into the coding equivalent of a black hole—*an infinite loop*. That is, writing a program that sooner or later somehow gets stuck in a never-ending circular execution of code and so, short of there being a power failure, never finishes. Wouldn't it be great to have a way to determine whether any program you had just written had (or didn't have) this flaw? This question, called appropriately enough, the *halting problem*, was examined by Turing, who showed that the halting problem has no solution. More precisely, imagine a computer program **H** that can (so you imagine), given as its inputs any computer program **P** and the input **I** to **P**, determine whether **P** when executing **I** will eventually halt (that is, not get trapped in an endless loop). This is illustrated in Figure A5.1.

Turing showed that the existence of **H** is impossible. That is, **H** is the computer science equivalent of a unicorn—you can certainly imagine it, yes, but it simply does not exist. Turing's reasoning is pretty abstract, but in 1952 the American mathematician Martin Davis (born 1928) gave a beautiful analysis that later appeared in a

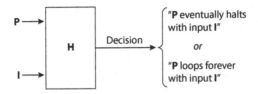

FIGURE A5.1.

H decides if **P**, executing **I**, either eventually halts or gets trapped forever
in an infinite loop.

college textbook[3] by the American mathematician Marvin Minsky (1927–2016). There Minsky wrote: "The result of the simple but delicate argument [in the proof of the insolvability of the halting problem] is perhaps the most important conclusion in this book." Davis' proof of Turing's result, as described in Minsky's book, takes the form of three steps.

Step 1: Any program **P** is ultimately described to the computer executing **H** as a sequence of 1s and 0s (in what is called *machine language*). The same is true for the input **I** to **P**, that is, **I** is also a sequential string of 1s and 0s. So, suppose we pick **I** to be **P**, that is, we present as input to **P** the same sequence of 1s and 0s that are the machine language version of **P**. Minsky writes of this perhaps curious choice for **I**: "We need not concern ourselves with the question of why anyone would be interested in such introverted calculations; still, there is nothing absurd about the notion of a man contemplating [with his brain] a description of his own brain." This choice for **I** does have some practical motivation that Minsky might have mentioned, however, because using **P** as its own input data automatically specifies the **I** we are going to use for any given **P**.

Step 2: Once we have **P** as the program input to **H**, along with **P** itself as the input **I** to **P**, we arrive at the situation shown in Figure A5.2, which shows **P** as the "double input" to **H**. **H** decides whether **P** (with input **P**) either eventually halts or loops forever. You'll

3. Marvin Minsky, *Computation: Finite and Infinite Machines* (Prentice-Hall, 1967). The quotes in the text are from pp. 148, 149.

notice that in Figure A5.2 we have introduced a bit of additional logic, logic that, once the halt/loop decision for **P** has been made, forces the final operation to be a halt if **H**'s decision was "**P** loops," or to be an infinite loop if **H**'s decision was "**P** halts." The presence of this extra logic is why everything inside the dashed box of Figure A5.2 is given a new name, **X**. Note that up to this point, we have *not* (yet) arrived at a paradoxical situation. But then comes . . .

Step 3: This last step is, as Minsky calls it, the "killer." Let **P** in Figure A5.2 be **X**. That is, apply **X** to itself. Figure A5.2 then becomes Figure A5.3, where we see that if **H** decides that **X** halts, then **X** loops; while if **H** decides that **X** loops, then **X** halts. Either way, **H** (with certainty) makes the *wrong* decision about **X**.

Minsky concludes his discussion of the halting problem with these words: "We have only the deepest sympathy for those readers who have not encountered this type of simple yet mind-boggling argument before. It resembles the argument in 'Russell's paradox' which forces us to discard arguments concerning the 'class of all

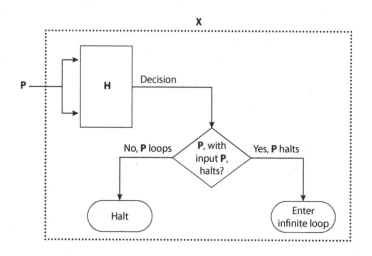

FIGURE A5.2.

Logic **X**.

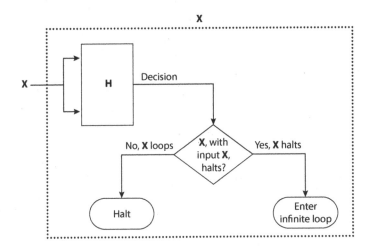

FIGURE A5.3.

X (an enhanced H) makes the wrong decision about itself.

classes'—a notion whose absurdity is not otherwise particularly evident." Minsky is referring to the English mathematician Bertrand Russell (1872–1970) and his famous puzzle of the village barber who shaves every man in the village who doesn't shave himself. The puzzle occurs when we ask: Who shaves the barber? The puzzle dates from 1902, and so Turing was clearly aware of it long before 1936.

In fact, Russell was not the first to see how a self-referencing condition can carry the seeds of its own destruction in the form of a derived self-contradiction. The earliest example of this is probably the ancient *liar's paradox* dating from four centuries before Christ. Ask yourself, after reading "This statement is a lie," whether it is true or not. It must be one or the other, right? Well, if it is a true statement (that is, it's a lie) then it's *not* true. And if it is a lie (that is, it's a true statement), then it's *not* a lie.

The fact that the ancients thought of a paradoxical situation that foreshadowed, by thousands of years, the modern-day computer science halting problem may make you think there is nothing left for modern analysts to tackle concerning computational questions. Not so. As I mentioned near the start of this appendix, Turing was greatly interested in the Riemann hypothesis, about which one of his biographers wrote "So far . . . no one [has] been able to prove [the Riemann hypothesis's] unprovability."[4]

To appreciate the challenge implicit in that statement, consider the following scene, in which we listen in on the conversation between two academic mathematicians as they enjoy a before-dinner glass of sherry in the faculty club dining room. Professors Tweedle and Twombly have just come from the weekly afternoon computer science seminar, where they listened to a guest speaker describe her calculation of 100 million zeros of the Riemann zeta function in a region of the complex plane vastly beyond anything explored before. To hardly anyone's surprise, every last one of those complex zeros was precisely on the critical line.

"I say, Twombly," says Tweedle, "that was a nice piece of computer programming by young Sweeny, but it really *proves* nothing at all. All it will take to disprove Riemann's conjecture, that all of the infinity of the zeta function's complex zeros are on the critical line, is finding just one zero that isn't. Now *that's* the real prize."

"You're right, Tweedle," replies Twombly, "although finding a theoretical *proof* that all the complex zeros *are* on the critical line, or perhaps not, would be just as good."

"Yes, yes," quickly agrees Tweedle, as he finishes off his sherry and pours another, "the RH question is a devilish one, isn't it? It's either true or it isn't, and glory to he or she who shows which it is."

"There *is*, however, one bright aspect to it, you know," says Twombly. "The question of the RH is, at least, not one of those damnable monsters that can be proven to be unsolvable."

"Oh, how's that, Twombly?" asks Tweedle, who is now on his third sherry.

4. See Leavitt, *The Man Who Knew Too Much*. The RH has proven to be such a difficult problem that some mathematicians have wondered if maybe the continuing failure to either prove or disprove it is because it is unsolvable. Such a situation would remove the embarrassing possibility that humans are just not smart enough!

"Well, look at it this way," answers Twombly, who is wondering if Tweedle is going to be able to follow the logic, with all that sherry in him. "If the RH could be shown to be unsolvable, that would mean nobody could ever experimentally find, using Sweeny's computer approach, a complex zero off the critical line, even by chance, no matter how long she looked. Not even if she could check zeros at the rate of $10^{10^{10}}$ each nanosecond. That's because if Sweeny did find such a rogue zero, then she would have proven the RH to be false, in contradiction with the initial premise there exists a proof that the question is unsolvable. But *that* would mean there *is* no zero off the critical line and that would mean the RH *had* been solved by showing it's true. That's a contradiction, too. The only way out of this quagmire is to conclude that no such unsolvability proof exists."

"Umph," grunts Tweedle, whose eyes appear to be looking in two different directions. A full minute of silence follows, and Twombly starts to suspect that the brilliance of his argument has hit a brick wall in the form of Tweedle's sherry-soaked brain. But he is wrong, as Tweedle suddenly (but carefully) puts his glass down and leans forward with a lopsided grin on his face.

"Well, Twombly," a now quite mellow Tweedle says, "I'm not so sure about all that. There are, after all, infinitely many complex zeros to the zeta function, and so no matter how many of them you check each nanosecond, it would take you infinite time to check them all. So, you see, you'd *never* be done with your calculations, and your contradictions don't occur *until* you are done. So, sorry old man, but I think your pretty little argument is pretty much a pretty big flop."

"Okay, Tweedle," grumbles Twombly, "maybe you're right. Or maybe you're not. Who knows? Anyway, here's a *real* problem for us to consider: What do we order for dinner? I hear that the fig-stuffed eggplant drenched in aged, probiotic yogurt is damn good, but so too, I'm told, is the high-fiber, fat-fried tofu burrito in a fermented snail sauce. What do you think?"

And with that we quickly tip-toe silently away, leaving the two old friends with a puzzle that makes the mystery of the RH seem, by comparison, utterly trivial.

Challenge Problem Solutions

1st Challenge in Preface: $I = \int_{-\infty}^{\infty} e^{-(ax^2-bx)}dx = \int_{-\infty}^{\infty} e^{-a(x^2-\frac{b}{a}x)}dx$

$= \int_{-\infty}^{\infty} e^{-a(x^2-\frac{b}{a}x+\frac{b^2}{4a^2}-\frac{b^2}{4a^2})}dx = \int_{-\infty}^{\infty} e^{-a(x^2-\frac{b}{a}x+\frac{b^2}{4a^2})}e^{\frac{b^2}{4a}}dx = e^{\frac{b^2}{4a}}\int_{-\infty}^{\infty} e^{-a(x-\frac{b}{2a})^2}dx$.

Let $u = x - \frac{b}{2a}$ and so $I = e^{\frac{b^2}{4a}}\int_{-\infty}^{\infty} e^{-au^2}du$. Let $u = \frac{s}{\sqrt{a}}$ and so

$I = e^{\frac{b^2}{4a}}\int_{-\infty}^{\infty} e^{-s^2}\frac{ds}{\sqrt{a}} = \frac{2}{\sqrt{a}}e^{\frac{b^2}{4a}}\int_{0}^{\infty} e^{-s^2}ds = \frac{2}{\sqrt{a}}e^{\frac{b^2}{4a}}\frac{1}{2}\sqrt{\pi} = \sqrt{\frac{\pi}{a}}e^{\frac{b^2}{4a}}$.

2nd Challenge in Preface: $\int_{a}^{b}\frac{dx}{\sqrt{(x-a)(b-x)}} = \int_{a}^{b}\frac{dx}{\sqrt{xb-x^2-ab+ax}}$

$= \int_{a}^{b}\frac{dx}{\sqrt{-ab+(a+b)x-x^2}}$. Notice that $\frac{(a-b)^2}{4}-\left(x-\frac{a+b}{2}\right)^2 = \frac{a^2-2ab+b^2}{4}$

$-[x^2-x(a+b)+\frac{(a+b)^2}{4}] = \frac{a^2-2ab+b^2}{4}-x^2+x(a+b)-\frac{a^2+2ab+b^2}{4}$

$= -ab+(a+b)x-x^2$. So, $\int_{a}^{b}\frac{dx}{\sqrt{(x-a)(b-x)}} = \int_{a}^{b}\frac{dx}{\sqrt{\frac{(a-b)^2}{4}-(x-\frac{a+b}{2})^2}}$. Let

$u = x - \frac{a+b}{2}$ (and so $du = dx$). Then $\int_{a}^{b}\frac{dx}{\sqrt{(x-a)(b-x)}} = \int_{\frac{a-b}{2}}^{\frac{b-a}{2}}\frac{du}{\sqrt{\frac{(a-b)^2}{4}-u^2}}$

$= \int_{\frac{b-a}{2}}^{\frac{b-a}{2}}\frac{du}{\sqrt{\frac{(b-a)^2}{4}-u^2}} = \int_{\frac{b-a}{2}}^{\frac{b-a}{2}}\frac{du}{\sqrt{\{\frac{b-a}{2}\}^2-u^2}} = \sin^{-1}\left(\frac{u}{\frac{b-a}{2}}\right)\Big|_{-\frac{b-a}{2}}^{\frac{b-a}{2}} = \sin^{-1}(1)-\sin^{-1}(-1)$

$= \frac{\pi}{2}-(-\frac{\pi}{2}) = \pi$, independent of a and b as long as $b > a \geq 0$.

3rd Challenge in Preface: The claim is obviously true for $q = 1$, as the left-hand side of the Botez-Catalan identity is $1-\frac{1}{2}=\frac{1}{2}$ while the right-hand side is $h(2)-h(1)=1+\frac{1}{2}-1=\frac{1}{2}$, too. Now, suppose the claim is true for $q = k$. That is, suppose that $1-\frac{1}{2}+\frac{1}{3}-\frac{1}{4}+\cdots-\frac{1}{2k}=h(2k)-h(k)$. If we can show that it then follows that the claim is true for $q = k + 1$, then the claim is true for all k. By our assumption, we have **(A)** $1-\frac{1}{2}+\frac{1}{3}-\frac{1}{4}+\cdots-\frac{1}{2k}+\frac{1}{2k+1}-\frac{1}{2k+2}$

$= h(2k) - h(k) + \frac{1}{2k+1} - \frac{1}{2k+2}$. Now, notice that $h\{2(k+1)\} - h\{k+1\}$ $= h\{2k+2\} - h\{k+1\}$ $= [h(2k) + \frac{1}{2k+1} + \frac{1}{2k+2}] - [h(k) + \frac{1}{k+1}] = h(2k) - h(k) + \frac{1}{2k+1}$ $+ \frac{1}{2k+2} - \frac{1}{k+1} = h(2k) - h(k) + \frac{1}{2k+1} + \frac{1}{2(k+1)} - \frac{1}{k+1} = h(2k)$ $- h(k) + \frac{1}{2k+1} - \frac{1}{2k+2}$, which is precisely the right-hand side of **(A)**. So, $h\{2(k+1)\} - k\{k+1\}$ must equal the left-hand side of **(A)**. That is, $h\{2(k+1)\} - k\{k+1\} = 1 - \frac{1}{2} + \frac{1}{3} - \frac{1}{4} + \cdots - \frac{1}{2(k+1)}$ and we are done.

Challenge Problem 1.1.1: Following the hint, $\frac{2^i + 2^{-i}}{2} = \frac{e^{\ln(2^i)} + e^{\ln(2^{-i})}}{2}$ $= \frac{e^{i\ln(2)} + e^{-i\ln(2)}}{2} = \frac{\cos\{\ln(2)\} + i\sin\{\ln(2)\} + \cos\{\ln(2)\} - i\sin\{\ln(2)\}}{2} = \cos\{\ln(2)\}$. It is a simple matter to numerically evaluate this, with the ready availability of electronic hand calculators (be sure, however, to set the calculator to its radian mode) to get $\cos\{\ln(2)\} = 0.769238\ldots$, while $\frac{10}{13} = 0.769230\ldots$. As Euler claimed, these two numbers are nearly equal, not differing until we get to the sixth decimal place.

Challenge Problem 1.2.1: Consider the geometric series $S = \sum_{k=1}^{\infty} \left(\frac{x}{2}\right)^k$ $= \left(\frac{x}{2}\right) + \left(\frac{x}{2}\right)^2 + \left(\frac{x}{2}\right)^3 + \left(\frac{x}{2}\right)^4 + \cdots$. Thus, $\left(\frac{x}{2}\right)S = \left(\frac{x}{2}\right)^2 + \left(\frac{x}{2}\right)^3 + \left(\frac{x}{2}\right)^4 + \cdots$, and so $S - \left(\frac{x}{2}\right)S = \frac{x}{2}$ or $S = \frac{\frac{x}{2}}{1 - \frac{x}{2}} = \frac{x}{2-x}$. Differentiating with respect to x, $\frac{dS}{dx} = \sum_{k=1}^{\infty} k \frac{x^{k-1}}{2^k}$, and so $\frac{dS}{dx}\big|_{x=1} = \sum_{k=1}^{\infty} \frac{k}{2^k}$. But $\frac{dS}{dx} = \frac{d}{dx}\{\frac{x}{2-x}\} = \frac{(2-x)-x(-1)}{(2-x)^2} = \frac{2}{(2-x)^2}$, and so $\frac{dS}{dx}\big|_{x=1} = 2$. That is, $\sum_{k=1}^{\infty} \frac{k}{2^k} = 2$. Differentiating again, $\frac{d^2S}{dx^2} = \sum_{k=1}^{\infty} k(k-1)\frac{x^{k-2}}{2^k} = \sum_{k=1}^{\infty} \frac{k^2 x^{k-2}}{2^k} - \sum_{k=1}^{\infty} \frac{kx^{k-2}}{2^k}$ or $\frac{d^2S}{dx^2}\big|_{x=1} = \sum_{k=1}^{\infty} \frac{k^2}{2^k} - \sum_{k=1}^{\infty} \frac{k}{2^k}$. But $\frac{d^2S}{dx^2} = \frac{d}{dx}\{\frac{2}{(2-x)^2}\} = \frac{-2[2(2-x)(-1)]}{(2-x)^4} = \frac{4}{(2-x)^3}$, and so $\frac{d^2S}{dx^2}\big|_{x=1} = 4$. So $4 = \sum_{k=1}^{\infty} \frac{k^2}{2^k} - 2$, or $\sum_{k=1}^{\infty} \frac{k^2}{2^k} = 6$. Now, differentiating once more, $\frac{d^3S}{dx^3} = \sum_{k=1}^{\infty} k(k-1)(k-2)\frac{x^{k-3}}{2^k} = \sum_{k=1}^{\infty} \frac{k^3 x^{k-3}}{2^k} - 3\sum_{k=1}^{\infty} \frac{k^2 x^{k-3}}{2^k} + 2\sum_{k=1}^{\infty} \frac{kx^{k-3}}{2^k}$, or $\frac{d^3S}{dx^3}\big|_{x=1} = \sum_{k=1}^{\infty} \frac{k^3}{2^k} - 3\sum_{k=1}^{\infty} \frac{k^2}{2^k} + 2\sum_{k=1}^{\infty} \frac{k}{2^k}$. But $\frac{d^3S}{dx^3} = \frac{d}{dx}\{\frac{4}{(2-x)^3}\} = \frac{-4[3(2-x)^2(-1)]}{(2-x)^6} = \frac{12}{(2-x)^4}$, and so $\frac{d^3S}{dx^3}\big|_{x=1} = 12$. Thus, $\sum_{k=1}^{\infty} \frac{k^3}{2^k} = 12 + 3(6) - 2(2) = 26$. Finally, differentiating yet again, $\frac{d^4S}{dx^4} = \sum_{k=1}^{\infty} k(k-1)(k-2)(k-3)\frac{x^{k-4}}{2^k} = \sum_{k=1}^{\infty} \frac{k^4 x^{k-4}}{2^k} - 6\sum_{k=1}^{\infty} \frac{k^3 x^{k-4}}{2^k}$ $+ 11\sum_{k=1}^{\infty} \frac{k^2 x^{k-4}}{2^k} - 6\sum_{k=1}^{\infty} \frac{kx^{k-4}}{2^k}$, or $\frac{d^4S}{dx^4}\big|_{x=1} = \sum_{k=1}^{\infty} \frac{k^4}{2^k} - 6\sum_{k=1}^{\infty} \frac{k^3}{2^k} + 11\sum_{k=1}^{\infty} \frac{k^2}{2^k} - 6\sum_{k=1}^{\infty} \frac{k}{2^k}$. But $\frac{d^4S}{dx^4}\big|_{x=1} = \frac{d}{dx}\{\frac{12}{(2-x)^4}\} = \frac{-12(4)(2-x)^3(-1)}{(2-x)^8} = \frac{48}{(2-x)^5}$, and so $\frac{d^4S}{dx^4}\big|_{x=1} = 48$. Thus, $\sum_{k=1}^{\infty} \frac{k^4}{2^k} = 48 + 6(26) - 11(6) + 6(2) \doteq 150$. Notice that these derivations show, as a side benefit, that Bernoulli's sums will be integer valued for all non-negative integer values of n in $\sum_{k=1}^{\infty} \frac{k^n}{2^k}$, which I don't think is a priori obvious. (The case of $n = 0$ can be directly summed to give $\sum_{k=1}^{\infty} \frac{1}{2^k} = 1$.) Note: It's easy to write *MATLAB* code (see **bersum.m** in the following box) to numerically compute Bernoulli's sums, and the theoretical values computed here are thereby confirmed. (The code is written as an endless loop, and to exit it requires the typing of a Control-C.) Just for fun, I ran the code for $n = 5$, and the code says $\sum_{k=1}^{\infty} \frac{k^5}{2^k} = 1{,}082$. If you've got some spare time on your hands, see if you can confirm this value theoretically.

```
%bersum.m
n=input('What is n?')
s=0;k=1;
while k>0
    s=s+(k^n)/(2^k)
    k=k+1;
end
```

Challenge Problem 1.3.1: $A = 1 + \frac{1}{3} + \frac{1}{5} + \frac{1}{7} + \cdots$ diverges to plus infinity, because it contains as a subset the reciprocals of all the primes (except for the first one), which diverges. $B = -\frac{1}{2} - \frac{1}{4} - \frac{1}{6} - \cdots = -\frac{1}{2}(1 + \frac{1}{2} + \frac{1}{3} + \cdots) = -\frac{1}{2}\zeta(1)$ diverges to minus infinity, because $\zeta(1) = \infty$. To arrange for the sum $A + B$ to converge to any desired value N, follow this procedure. Start by adding the terms of A until the partial sum first exceeds N. Then switch to adding in the terms of B until the partial sum first falls below N. Then switch back to A until the partial sum again exceeds N. Then switch back to B, and so on.

Challenge Problem 1.4.1: Setting $n = \frac{3}{2}$ in (1.4.1), we have $\Gamma(\frac{3}{2}) = \int_0^{\infty} e^{-x} x^{\frac{1}{2}} dx = (\frac{1}{2})!$ by (1.4.4). Integrating by parts, with $u = x^{\frac{1}{2}}$ and $dv = e^{-x} dx$, we have $v = -e^{-x}$ and $\frac{du}{dx} = \frac{1}{2}x^{-\frac{1}{2}} = \frac{1}{2x^{\frac{1}{2}}}$ and so $du = \frac{1}{2x^{\frac{1}{2}}} dx$. Thus, $\int_0^{\infty} e^{-x} x^{\frac{1}{2}} dx = (-x^{\frac{1}{2}}e^{-x})|_0^{\infty} + \int_0^{\infty} \frac{e^{-x}}{2x^{\frac{1}{2}}} dx = \frac{1}{2}\int_0^{\infty} \frac{e^{-x}}{x^{\frac{1}{2}}} dx = \frac{1}{2}\Gamma(\frac{1}{2})$ or, using (1.4.6), $(\frac{1}{2})! = \frac{1}{2}\sqrt{\pi}$.

Challenge Problem 1.4.2: For n any positive integer, $n!$ is a finite (positive) integer. By (1.4.19) we have $(-n)! = \frac{n\pi}{n!\sin(n\pi)}$, which blows up because $\sin(n\pi) = 0$. The direction of the blow-up depends on whether n is approached from below or above, but in any case, the absolute value of $(-n)!$ is infinity for n any positive integer.

Challenge Problem 1.5.1: Let $u = e^{-x}$ (and so $e^x = \frac{1}{u}$). Thus, $\frac{du}{dx} = -e^{-x} = -u$ or $dx = -\frac{du}{u}$. Therefore, $\int_{-\infty}^{\infty} \frac{e^{px}}{1+e^x} dx = \int_{\infty}^{0} \frac{\frac{1}{u^p}}{1+\frac{1}{u}} (-\frac{du}{u})$ $= \int_{0}^{\infty} \frac{\frac{1}{u^p}}{u+1} (\frac{du}{u}) = \int_{0}^{\infty} \frac{du}{(1+u)u^p}$, which is the integral in (1.5.10).

Challenge Problem 1.6.1: If we had made the substitution of $e^{-x} = -\frac{d}{dx}(e^{-x} - C)$, we would have arrived at (1.6.4) with the integral $\int_{0}^{1} \frac{C - e^{-x}}{x} dx$, an integral that exists only if $C = 1$. Here's how to see this. The integrand is $\frac{C - e^{-x}}{x} = \frac{C - \{1 - x + \frac{x^2}{2!} - \frac{x^3}{3!} + \frac{x^4}{4!} - \cdots\}}{x} = \frac{C-1}{x} + 1 - \frac{x}{2!} + \frac{x^2}{3!} - \frac{x^3}{4!} + \cdots$ and so $\int_{0}^{1} \frac{C - e^{-x}}{x} dx = \int_{0}^{1} \{\frac{C-1}{x} + 1 - \frac{x}{2!} + \frac{x^2}{3!} - \frac{x^3}{4!} + \cdots\} dx = \{(C-1)\ln(x) + x - \frac{x^2}{2(2!)} + \frac{x^3}{3(3!)}$ $-\frac{x^4}{4(4!)} + \cdots\}|_{0}^{1}$, which in general blows up at the lower limit because of the log term. For the lone case of $C = 1$, however, the log term disappears, and the integral is well behaved: $\int_{0}^{1} \frac{1 - e^{-x}}{x} dx = 1 - \frac{1}{2(2!)} + \frac{1}{3(3!)} - \frac{1}{4(4!)} + \cdots$, which clearly converges to a finite value (see note 6 in Chapter 1).

Challenge Problem 1.6.2: Following the hint, put $n = 0$ into (1.6.7) to get $\int_{0}^{1} \{\frac{1}{x}\} dx = \sum_{k=1}^{\infty} \int_{k}^{k+1} \frac{dy}{y} - \sum_{k=1}^{\infty} k \int_{k}^{k+1} \frac{dy}{y^2}$. Now, $\int_{k}^{k+1} \frac{dy}{y} = \ln(y)|_{k}^{k+1}$ $= \ln(k+1) - \ln(k)$ and $\int_{k}^{k+1} \frac{dy}{y^2} = -y^{-1}|_{k}^{k+1} = -[\frac{1}{k+1} - \frac{1}{k}] = \frac{1}{k} - \frac{1}{k+1} = \frac{1}{k(k+1)}$. So

$$\int_{0}^{1} \{\frac{1}{x}\} dx = \lim_{q \to \infty} \{\sum_{k=1}^{q} [\ln(k+1) - \ln(k)] - \sum_{k=1}^{q} k \frac{1}{k(k+1)}\}$$

$$= \lim_{q \to \infty} \{\sum_{k=1}^{q} [\ln(k+1) - \ln(k)] - \sum_{k=1}^{q} \frac{1}{(k+1)}\}$$

$$= \lim_{q \to \infty} \{[\ln(2) - \ln(1)] + [\ln(3) - \ln(2)] + \cdots + [\ln(q+1) - \ln(q)]\}$$

$$-\lim_{q \to \infty} (\frac{1}{2} + \frac{1}{3} + \cdots + \frac{1}{q+1}) = \lim_{q \to \infty} \ln(q+1) - \lim_{q \to \infty} [(1 + \frac{1}{2} + \frac{1}{3} + \cdots + \frac{1}{q+1}) - 1]$$

$$= \lim_{q \to \infty} [\ln(q+1) - h(q+1) + 1] = \lim_{q \to \infty} [-\{h(q+1) - \ln(q+1)\} + 1] = -\gamma + 1.$$

Challenge Problem 1.6.3: Following the hint, let $1 - x = e^{-t}$ and so $-\frac{dx}{dt} = -e^{-t}$ or $dx = e^{-t} dt$. Thus, $\int_{0}^{1} \frac{\{\ln(1-x)\}^2}{x} dx = \int_{0}^{\infty} \frac{\{\ln(e^{-t})\}^2}{1 - e^{-t}} e^{-t} dt$ $= \int_{0}^{\infty} \frac{\{-t\}^2}{1 - e^{-t}} e^{-t} dt = \int_{0}^{\infty} t^2 \frac{e^{-t}}{1 - e^{-t}} dt = \int_{0}^{\infty} \frac{t^2}{e^t - 1} dt$. This is the $s = 3$ case in (1.4.24), which says $\int_{0}^{\infty} \frac{x^2}{e^x - 1} dx = \Gamma(3)\zeta(3) = 2!\zeta(3) = 2\zeta(3)$.

Challenge Problem 1.7.1: $\Gamma(x) = \int_0^\infty e^{-t} t^{x-1} dt$ and so $\Gamma(\frac{1}{n}) = \int_0^\infty e^{-t} t^{\frac{1}{n}-1} dt$ $= \int_0^\infty \frac{e^{-t}}{t^{\frac{n-1}{n}}} dt$. Changing variable to $t = u^n$ ($dt = nu^{n-1} du$), we have $\Gamma(\frac{1}{n}) = \int_0^\infty \frac{e^{-u^n}}{(u^n)^{\frac{n-1}{n}}} nu^{n-1} du = n \int_0^\infty e^{-u^n} du$ or recalling that $x\Gamma(x) = x!$, we have $\frac{1}{n} \Gamma(\frac{1}{n}) = \int_0^\infty e^{-u^n} du = (\frac{1}{n})!$

Challenge Problem 1.7.2: Setting $x = 1$ in (1.7.20)/(1.7.21) gives $\psi(1) = -1 - \gamma + \Sigma_{k=1}^\infty \{\frac{1}{k} - \frac{1}{k+1}\}$. Since $\Sigma_{k=1}^\infty \{\frac{1}{k} - \frac{1}{k+1}\} = \{\frac{1}{1} - \frac{1}{2}\} + \{\frac{1}{2} - \frac{1}{3}\} + \{\frac{1}{3} - \frac{1}{4}\} + \cdots = 1$, we immediately have $\psi(1) = \frac{\Gamma'(x)}{\Gamma(x)}\big|_{x=1} = -\gamma$.

Challenge Problem 1.7.3: Following the hint, write $I(m) = \int_0^\infty x^m e^{-x} dx = \int_0^\infty e^{m \ln(x)} e^{-x} dx$. Then $\frac{dI}{dm} = \int_0^\infty \ln(x) e^{m \ln(x)} e^{-x} dx$, and so $\frac{d^2 I}{dm^2} = \int_0^\infty \ln^2(x) e^{m \ln(x)} e^{-x} dx$. Thus, $\int_0^\infty e^{-x} \ln^2(x) dx = \frac{d^2 I}{dm^2}\big|_{m=0}$. Notice that $I(m)$ is the gamma function $\Gamma(n)$ for $n - 1 = m$, that is, $n = m + 1$. So, $I(m) = \Gamma(m+1)$ and $\int_0^\infty e^{-x} \ln^2(x) dx = \{\frac{d^2}{dm^2} \Gamma(m+1)\}\big|_{m=0}$. Now, the digamma function says $\frac{d\Gamma(z)}{dz} = \Gamma(z)[-\frac{1}{z} - \gamma + \Sigma_{r=1}^\infty (\frac{1}{r} - \frac{1}{r+z})]$ and so, for $z = m+1$, $\frac{d\Gamma(m+1)}{d(m+1)} = \frac{d\Gamma(m+1)}{dm} = \Gamma(m+1)[-\frac{1}{m+1} - \gamma + \Sigma_{r=1}^\infty (\frac{1}{r} - \frac{1}{r+m+1})]$. Differentiating again, $\frac{d^2 \Gamma(m+1)}{dm^2} = \frac{d\Gamma(m+1)}{dm}[-\frac{1}{m+1} - \gamma + \Sigma_{r=1}^\infty (\frac{1}{r} - \frac{1}{r+m+1})] + \Gamma(m+1)[\frac{1}{(m+1)^2} + \Sigma_{r=1}^\infty \frac{1}{(r+m+1)^2}]$. So since $\Gamma(1) = 1$, $\frac{d\Gamma(m+1)}{dm}\big|_{m=0} = \Gamma(1)[-1 - \gamma + \Sigma_{r=1}^\infty (\frac{1}{r} - \frac{1}{r+1})]$ $= -1 - \gamma + (1 - \frac{1}{2}) + (\frac{1}{2} - \frac{1}{3}) + \cdots = -1 - \gamma + 1 = -\gamma$ and $\frac{d^2 \Gamma(m+1)}{dm^2}\big|_{m=0} = -\gamma[-1 - \gamma + \Sigma_{r=1}^\infty (\frac{1}{r} - \frac{1}{r+1})] + \Gamma(1)[1 + \Sigma_{r=1}^\infty \frac{1}{(r+1)^2}] = \gamma + \gamma^2 - \gamma[(1 - \frac{1}{2}) + (\frac{1}{2} - \frac{1}{3}) + \cdots] + 1 + \frac{1}{2^2} + \frac{1}{3^2} + \cdots = \gamma + \gamma^2 - \gamma + 1 + \frac{1}{2^2} + \frac{1}{3^2} + \cdots = \gamma^2 + \zeta(2) = \gamma^2 + \frac{\pi^2}{6} = \int_0^\infty e^{-x} \ln^2(x) dx$.

Challenge Problem 1.7.4: Following the hint, $\Gamma(x+1) = e^{-\gamma x + \Sigma_{k=2}^\infty (-1)^k \frac{\zeta(k)}{k} x^k}$ $= e^{-\gamma x} e^{\Sigma_{k=2}^\infty (-1)^k \frac{\zeta(k)}{k} x^k}$ $= [1 - \gamma x + \frac{\gamma^2}{2!} x^2 - \frac{\gamma^3}{3!} x^3 + \frac{\gamma^4}{4!} x^4 - \cdots]$ $e^{\frac{\zeta(2)}{2} x^2} e^{-\frac{\zeta(3)}{3} x^3} e^{\frac{\zeta(4)}{4} x^4} \cdots$ $= (1 - \gamma x + \frac{\gamma^2}{2} x^2 - \frac{\gamma^3}{6} x^3 + \frac{\gamma^4}{24} x^4 + \cdots) (1 + \frac{\zeta(2)}{2} x^2 + \frac{\zeta^2(2)}{8} x^4 + \cdots) (1 - \frac{\zeta(3)}{3} x^3 + \cdots)$ $(1 + \frac{\zeta(4)}{4} x^4 + \cdots) \cdots = 1 - \gamma x + [\frac{\gamma^2}{2} + \frac{\zeta(2)}{2}] x^2 + [-\frac{\gamma^3}{6} - \frac{\zeta(3)}{3} - \frac{\gamma\zeta(2)}{2}] x^3 + \cdots$. Thus, the coefficient of x^3 is $[-\frac{\gamma^3}{6} - \frac{\zeta(3)}{3} - \frac{\gamma\pi^2}{12}] = -\frac{1}{3}[\frac{1}{2}\gamma^3 + \zeta(3) + \frac{1}{4}\gamma\pi^2]$. In the same way, looking at all the possible ways to form terms in x^4, we find that the coefficient of x^4 is $\frac{\gamma^4}{24} + \frac{\zeta^2(2)}{8} + \frac{\zeta(4)}{4} + \frac{\gamma^2\zeta(2)}{4} + \frac{\gamma\zeta(3)}{3}$.

Challenge Problem 1.7.5: $i^i = (e^{i\frac{\pi}{2}})^i = e^{i^2 \frac{\pi}{2}} = e^{-\frac{\pi}{2}} = \frac{1}{e^{\pi/2}} = \frac{1}{\sqrt{e^\pi}} = 0.2078795 \ldots$

Challenge Problem 1.8.1: Following the hint, we have $\frac{1}{4n(4n-1)(4n+1)}$ $= \frac{A}{4n} + \frac{B}{4n-1} + \frac{C}{4n+1}$. To find A, multiply through by $4n$ and then set $n = 0$ to get $A = \frac{1}{(4n-1)(4n+1)}\big|_{n=0} = -1$. To find B, multiply through by $4n - 1$ and

then set $n=\frac{1}{4}$ to get $B=\frac{1}{4n(4n+1)}\big|_{n=\frac{1}{4}}=\frac{1}{2}$. To find C, multiply through by $4n+1$ and then set $n=-\frac{1}{4}$ to get $C=\frac{1}{4n(4n-1)}\big|_{n=-\frac{1}{4}}=\frac{1}{2}$. Thus, $\frac{1}{4n(4n-1)(4n+1)}$
$=-\frac{1}{4n}+\frac{\frac{1}{2}}{4n-1}+\frac{\frac{1}{2}}{4n+1}=\frac{1}{2}(\frac{1}{4n-1}+\frac{1}{4n+1}-\frac{1}{2n})$ and so $\sum_{n=1}^{\infty}\frac{1}{(4n)^3-4n}=\frac{1}{2}\sum_{n=1}^{\infty}(\frac{1}{4n-1}+\frac{1}{4n+1}-\frac{1}{4n}$
$-\frac{1}{4n}+\frac{1}{4n+2}-\frac{1}{4n+2})$ where the (perhaps) non-obvious steps of writing $\frac{1}{2n}=\frac{1}{4n}+\frac{1}{4n}$ and of adding and subtracting $\frac{1}{4n+2}$ have been done.
Regrouping, $\sum_{n=1}^{\infty}\frac{1}{(4n)^3-4n}=\frac{1}{2}\sum_{n=1}^{\infty}(\frac{1}{4n-1}-\frac{1}{4n}+\frac{1}{4n+1}-\frac{1}{4n+2})-\frac{1}{2}\sum_{n=1}^{\infty}(\frac{1}{4n}-\frac{1}{4n+2})$
$=\frac{1}{2}\sum_{n=1}^{\infty}(\frac{1}{4n-1}-\frac{1}{4n}+\frac{1}{4n+1}-\frac{1}{4n+2})-\frac{1}{4}\sum_{n=1}^{\infty}(\frac{1}{2n}-\frac{1}{2n+1})$. Now, $\sum_{n=1}^{\infty}(\frac{1}{4n-1}-\frac{1}{4n}+\frac{1}{4n+1}-\frac{1}{4n+2})$
$=(\frac{1}{3}-\frac{1}{4}+\frac{1}{5}-\frac{1}{6})+(\frac{1}{7}-\frac{1}{8}+\frac{1}{9}-\frac{1}{10})+\cdots$ where on the right the first pair of parentheses is for $n=1$, the second pair is for $n=2$, and so on. That is, $\sum_{n=1}^{\infty}(\frac{1}{4n-1}-\frac{1}{4n}+\frac{1}{4n+1}-\frac{1}{4n+2})=\frac{1}{3}-\frac{1}{4}+\frac{1}{5}-\frac{1}{6}+\frac{1}{7}-\frac{1}{8}+\frac{1}{9}-\frac{1}{10}+\cdots$. From (1.3.5), with $x=1$, we have $\ln(2)=1-\frac{1}{2}+\frac{1}{3}-\frac{1}{4}+\frac{1}{5}-\cdots$ and so $\frac{1}{3}-\frac{1}{4}+\frac{1}{5}-\cdots$
$=\ln(2)-1+\frac{1}{2}=\ln(2)-\frac{1}{2}$. Thus, $\sum_{n=1}^{\infty}\frac{1}{(4n)^3-4n}=\frac{1}{2}\ln(2)-\frac{1}{4}-\frac{1}{4}\sum_{n=1}^{\infty}(\frac{1}{2n}-\frac{1}{2n+1})$.
Now, $\sum_{n=1}^{\infty}(\frac{1}{2n}-\frac{1}{2n+1})=(\frac{1}{2}-\frac{1}{3})+(\frac{1}{4}-\frac{1}{5})+(\frac{1}{6}-\frac{1}{7})+\cdots=\frac{1}{2}-\frac{1}{3}+\frac{1}{4}-\frac{1}{5}+\frac{1}{6}-\frac{1}{7}+\cdots$.
From before, we have $-\ln(2)=-1+\frac{1}{2}-\frac{1}{3}+\frac{1}{4}-\frac{1}{5}+\cdots$ and so $\frac{1}{2}-\frac{1}{3}+\frac{1}{4}-\frac{1}{5}+\cdots$
$=1-\ln(2)$. So $\sum_{n=1}^{\infty}\frac{1}{(4n)^3-4n}=\frac{1}{2}\ln(2)-\frac{1}{4}-\frac{1}{4}[1-\ln(2)]=-\frac{1}{2}+\frac{3}{4}\ln(2)$. Thus, $1+2\sum_{n=1}^{\infty}\frac{1}{(4n)^3-4n}=1+2[-\frac{1}{2}+\frac{3}{4}\ln(2)]=1-1+\frac{3}{2}\ln(2)=\frac{3}{2}\ln(2)$, as was to be shown.

Challenge Problem 1.8.2: Making the partial fraction expansion $\frac{1}{n(n+1)(n+2)(n+3)}=\frac{A}{n}+\frac{B}{n+1}+\frac{C}{n+2}+\frac{D}{n+3}$ and using the same procedure as in the text (and previous problem), we find that $\sum_{n=1}^{\infty}\frac{1}{n(n+1)(n+2)(n+3)}$
$=\sum_{n=1}^{\infty}\{\frac{1/6}{n}-\frac{\frac{1}{2}}{n+1}+\frac{\frac{1}{2}}{n+2}-\frac{1/6}{n+3}\}=\frac{1}{6}\sum_{n=1}^{\infty}\{\frac{1}{n}-\frac{3}{n+1}+\frac{3}{n+2}-\frac{1}{n+3}\}=\frac{1}{6}[(1-\frac{3}{2}+\frac{3}{3}-\frac{1}{4})$
$+(\frac{1}{2}-\frac{3}{3}+\frac{3}{4}-\frac{1}{5})+(\frac{1}{3}-\frac{3}{4}+\frac{3}{5}-\frac{1}{6})+(\frac{1}{4}-\frac{3}{5}+\frac{3}{6}-\frac{1}{7})+(\frac{1}{5}-\frac{3}{6}+\frac{3}{7}-\frac{1}{8})+(\frac{1}{6}-\frac{3}{7}+\frac{3}{8}-\frac{1}{9})$
$+(\frac{1}{7}-\frac{3}{8}+\frac{3}{9}-\frac{1}{10})+\cdots]$, and you can see that the series telescopes, with only the first term in the third pair of parentheses ($\frac{1}{3}$) surviving. So $\sum_{n=1}^{\infty}\frac{1}{n(n+1)(n+2)(n+3)}=\frac{1}{18}$.

Challenge Problem 1.8.3: See the end of Chapter 2 for a detailed discussion.

Challenge Problem 2.1.1: Following the hint, we have $k!=\int_0^{\infty}x^k e^{-x}dx$, and so $S=\sum_{k=0}^{\infty}(-1)^k\int_0^{\infty}x^k e^{-x}dx=\int_0^{\infty}e^{-x}\sum_{k=0}^{\infty}(-1)^k x^k dx$
$=\int_0^{\infty}e^{-x}\{1-x+x^2-x^3+\cdots\}dx=\int_0^{\infty}\frac{e^{-x}}{1+x}dx<\int_0^{\infty}e^{-x}dx=(-e^{-x})|_0^{\infty}=1$. Thus, S is

finite. Euler numerically evaluated[1] $\int_0^\infty \frac{e^{-x}}{1+x}dx$ using the trapezoidal rule to get a more precise value for S of about 0.59, and I used computer software (*MATLAB*) to get a value of 0.59634 . . . .

Challenge Problem 2.1.2: Following Poisson, write $T(x) = 1 - 2x + 4x^2 - 8x^3 + 16x^4 - 32x^5 + \ldots$ or $T(x) = 1 - (2x) + (2x)^2 - (2x)^3 + (2x)^4 - (2x)^5 + \cdots$, and so $(2x)T(x) = (2x) - (2x)^2 + (2x)^3 - (2x)^4 + (2x)^5 - \cdots$. Therefore, adding these two expressions, we have $T(x) + (2x)T(x) = 1$, or $T(x) = \frac{1}{1+2x}$ and so $\lim_{x \to 1} T(x) = 1 - 2 + 4 - 8 + 16 - 32 + \cdots = \frac{1}{3}$.

Challenge Problem 2.1.3: From the text, $\eta(s) = (1 - 2^{1-s})\zeta(s)$, and so setting $s = 0$, $\eta(0) = (1 - 2^1)\zeta(0) = -\zeta(0)$. Since $\eta(0) = \frac{1}{1^0} - \frac{1}{2^0} + \frac{1}{3^0} - \frac{1}{4^0} + \cdots = 1 - 1 + 1 - 1 + \cdots$, which is $\frac{1}{2}$ by (2.1.1), then $\zeta(0) = -\frac{1}{2}$. Setting $s = -2$, $\eta(-2) = (1 - 2^3)\zeta(-2) = -7\zeta(-2)$, or $\zeta(-2) = -\frac{1}{7}\eta(-2)$. Since $\eta(-2) = \frac{1}{1^{-2}} - \frac{1}{2^{-2}} + \frac{1}{3^{-2}} - \frac{1}{4^{-2}} + \cdots = 1 - 2^2 + 3^2 - 4^2 + \cdots = 0$ by (2.1.3), then $\zeta(-2) = 0$. (Notice that this is consistent with the claim made in the first box of Section 1.2, that all the even negative integers are zeros of the zeta function.) And setting $s = -3$, $\eta(-3) = (1 - 2^4)\zeta(-3) = -15\zeta(-3)$, or $\zeta(-3) = -\frac{1}{15}\eta(-3)$. Since $\eta(-3) = \frac{1}{1^{-3}} - \frac{1}{2^{-3}} + \frac{1}{3^{-3}} - \frac{1}{4^{-3}} + \cdots = 1 - 2^3 + 3^3 - 4^3 + \cdots = -\frac{1}{8}$ by (2.1.4), then $\zeta(-3) = -\frac{1}{15}(-\frac{1}{8}) = \frac{1}{120}$. Finally, setting $s = \frac{1}{2}$, we have $\eta(\frac{1}{2}) = (1 - 2^{1-\frac{1}{2}})\zeta(\frac{1}{2}) = (1 - \sqrt{2})\zeta(\frac{1}{2})$. Since $\eta(\frac{1}{2}) = \sum_{k=1}^\infty \frac{(-1)^{k+1}}{k^{1/2}} = 1 - \frac{1}{\sqrt{2}} + \frac{1}{\sqrt{3}} - \frac{1}{\sqrt{4}} + \cdots$, then $\zeta(\frac{1}{2}) = \frac{1}{1-\sqrt{2}}[1 - \frac{1}{\sqrt{2}} + \frac{1}{\sqrt{3}} - \frac{1}{\sqrt{4}} + \cdots]$, which we know converges (by note 6 in Chapter 1). I evaluated this expression using the first 1 million terms of the alternating series, which gives $\zeta(\frac{1}{2}) \approx -1.459147 \ldots$. The same theorem that tells us the alternating series converges also says that the maximum error we make in using a finite number of terms is less than the first term neglected, and so our estimate has an error with magnitude less than $|\frac{1}{1-\sqrt{2}}(\frac{1}{\sqrt{1,000,000}})| \approx 0.0025$. In fact, the true value of $\zeta(\frac{1}{2})$ is $-1.46035 \ldots$, and so our actual error is only about 0.0012.

1. For the details of Euler's numerical computations, see E. J. Burbeau, "Euler Subdues a Very Obstreperous Series," *American Mathematical Monthly*, May 1979, pp. 356–372.

Challenge Problem 2.2.1: Using $(n-1)! = \frac{n!}{n}$ and starting with $n = -4\frac{1}{2}$, we have

$$(-5\tfrac{1}{2})! = \frac{(-4\frac{1}{2})!}{-4\frac{1}{2}} = -\tfrac{2}{9}(-4\tfrac{1}{2})!,$$

$$(-4\tfrac{1}{2})! = \frac{(-3\frac{1}{2})!}{-3\frac{1}{2}} = -\tfrac{2}{7}(-3\tfrac{1}{2})!,$$

$$(-3\tfrac{1}{2})! = \frac{(-2\frac{1}{2})!}{-2\frac{1}{2}} = -\tfrac{2}{5}(-2\tfrac{1}{2})!,$$

$$(-2\tfrac{1}{2})! = \frac{(-1\frac{1}{2})!}{-1\frac{1}{2}} = -\tfrac{2}{3}(-1\tfrac{1}{2})!,$$

$$(-1\tfrac{1}{2})! = \frac{(-\frac{1}{2})!}{-\frac{1}{2}} = -2\left(-\tfrac{1}{2}\right)! = -2\sqrt{\pi}$$

where $(-\frac{1}{2})! = \sqrt{\pi}$ is our result from (1.4.6). Thus,

$$(-5\tfrac{1}{2})! = (-\tfrac{2}{9})(-\tfrac{2}{7})(-\tfrac{2}{5})(-\tfrac{2}{3})(-2\sqrt{\pi}) = -\tfrac{2^5}{945}\sqrt{\pi} = -0.0600196\ldots.$$

Challenge Problem 2.3.1: Starting with the defining series for $\zeta(2)$, we have $\zeta(2) = \frac{1}{1^2} + \frac{1}{2^2} + \frac{1}{3^2} + \frac{1}{4^2} + \frac{1}{5^2} + \frac{1}{6^2} + \cdots = \{\frac{1}{1^2} + \frac{1}{3^2} + \frac{1}{5^2} + \cdots\} + \{\frac{1}{2^2} + \frac{1}{4^2} + \frac{1}{6^2} + \cdots\}$ $= \{\frac{1}{1^2} + \frac{1}{3^2} + \frac{1}{5^2} + \cdots\} + \{\frac{1}{(2\times1)^2} + \frac{1}{(2\times2)^2} + \frac{1}{(2\times3)^2} + \cdots\} = \{\frac{1}{1^2} + \frac{1}{3^2} + \frac{1}{5^2} + \cdots\} + \frac{1}{4}\{\frac{1}{1^2} + \frac{1}{2^2} + \frac{1}{3^2} + \cdots\}$ $= \{\frac{1}{1^2} + \frac{1}{3^2} + \frac{1}{5^2} + \cdots\} + \frac{1}{4}\zeta(2)$ and so $\frac{1}{1^2} + \frac{1}{3^2} + \frac{1}{5^2} + \cdots = \frac{3}{4}\zeta(2)$. Thus, looking back at the start of this chain, $\zeta(2) = \frac{3}{4}\zeta(2) + \{\frac{1}{2^2} + \frac{1}{4^2} + \frac{1}{6^2} + \cdots\}$, or $\frac{1}{2^2} + \frac{1}{4^2} + \frac{1}{6^2} + \cdots = \frac{1}{4}\zeta(2)$ and we have our first result. To get the second result, start with our earlier result $\frac{1}{1^2} + \frac{1}{3^2} + \frac{1}{5^2} + \cdots = \frac{3}{4}\zeta(2)$, which says that $\frac{2}{1^2} + \frac{2}{3^2} + \frac{2}{5^2} + \cdots = \frac{3}{2}\zeta(2)$. From this subtract $\frac{1}{1^2} + \frac{1}{2^2} + \frac{1}{3^2} + \frac{1}{4^2} + \frac{1}{5^2} + \cdots = \zeta(2)$ to get $\frac{1}{1^2} - \frac{1}{2^2} + \frac{1}{3^2} - \frac{1}{4^2} + \frac{1}{5^2} - \cdots = \frac{1}{2}\zeta(2)$, and we are done.

Challenge Problem 2.4.1: Recalling the definitions of P and Q, $P = \int_0^1 \int_0^1 \frac{dxdy}{1-xy}$ and $Q = \int_0^1 \int_0^1 \frac{dxdy}{1+xy}$, we have $P + Q = \int_0^1 \int_0^1 \{\frac{1}{1-xy} + \frac{1}{1+xy}\} dxdy$ $= 2\int_0^1 \int_0^1 \frac{dxdy}{1-x^2y^2}$. Since $P = \frac{\pi^2}{6}$ and $Q = \frac{\pi^2}{12}$, then $\int_0^1 \int_0^1 \frac{dxdy}{1-x^2y^2} = \frac{1}{2}(\frac{\pi^2}{6} + \frac{\pi^2}{12}) = \frac{\pi^2}{8}$.

Challenge Problem 2.4.2: We have $\frac{x^n y^n}{1-xy} = x^n y^n[1 + xy + (xy)^2 + (xy)^3 + (xy)^4 + \cdots] = x^n y^n + x^{n+1} y^{n+1} + x^{n+2} y^{n+2} + x^{n+3} y^{n+3} + x^{n+4} y^{n+4} + \cdots$ and so

$$\int_0^1 \int_0^1 \frac{x^n y^n}{1-xy} dxdy = \int_0^1 \{\int_0^1 (x^n y^n + x^{n+1}y^{n+1} + x^{n+2}y^{n+2} + x^{n+3}y^{n+3} + \cdots) dx\} dy$$

$$= \int_0^1 (\tfrac{1}{n+1} x^{n+1} y^n + \tfrac{1}{n+2} x^{n+2} y^{n+1} + \tfrac{1}{n+3} x^{n+3} y^{n+2} + \tfrac{1}{n+4} x^{n+4} y^{n+3} + \cdots)\big|_0^1 \, dy$$

$$= \int_0^1 (\tfrac{1}{n+1} y^n + \tfrac{1}{n+2} y^{n+1} + \tfrac{1}{n+3} y^{n+2} + \tfrac{1}{n+4} y^{n+3} + \cdots) dy$$

$$= (\tfrac{1}{n+1}\tfrac{1}{n+1} y^{n+1} + \tfrac{1}{n+2}\tfrac{1}{n+2} y^{n+2} + \tfrac{1}{n+3}\tfrac{1}{n+3} y^{n+3} + \cdots)\big|_0^1$$

$$= \frac{1}{(n+1)^2} + \frac{1}{(n+2)^2} + \frac{1}{(n+3)^2} + \cdots = \{\tfrac{1}{1^2} + \tfrac{1}{2^2} + \tfrac{1}{3^2} + \cdots + \tfrac{1}{n^2}\} + \{\tfrac{1}{(n+1)^2} + \tfrac{1}{(n+2)^2} + \tfrac{1}{(n+3)^2} + \cdots\}$$

$$-\{\tfrac{1}{1^2} + \tfrac{1}{2^2} + \tfrac{1}{3^2} + \cdots + \tfrac{1}{n^2}\} = \zeta(2) - \{\tfrac{1}{1^2} + \tfrac{1}{2^2} + \tfrac{1}{3^2} + \cdots + \tfrac{1}{n^2}\}. \text{ If } n=3, \zeta(2) - \{\tfrac{1}{1^2} + \tfrac{1}{2^2} + \tfrac{1}{3^2}\}$$

$$= \zeta(2) - \{1 + \tfrac{1}{4} + \tfrac{1}{9}\} = \zeta(2) - \{\tfrac{36}{36} + \tfrac{9}{36} + \tfrac{4}{36}\} = \zeta(2) - \tfrac{49}{36}, \text{ or } \int_0^1 \int_0^1 \frac{x^3 y^3}{1-xy} dxdy = \zeta(2) - (\tfrac{7}{6})^2.$$

Challenge Problem 2.4.3: Define $P = \int_0^1 \int_0^1 \int_0^1 \frac{dxdydz}{1-xyz}$ and $= \int_0^1 \int_0^1 \int_0^1 \frac{dxdydz}{1+xyz}$. Then $P - Q = \int_0^1 \int_0^1 \int_0^1 \{\frac{1}{1-xyz} - \frac{1}{1+xyz}\} dxdydz = 2\int_0^1 \int_0^1 \int_0^1 \frac{xyz}{1-x^2y^2z^2} dxdydz$. Let $u = x^2, v = y^2, w = z^2$. Then $dxdydz = \frac{dudvdw}{8xyz}$ and $P - Q = \frac{1}{4}\int_0^1 \int_0^1 \int_0^1 \frac{dudvdw}{1-uvw} = \frac{1}{4}P$, or $Q = \frac{3}{4}P$. Now, recall from note 17 in Chapter 2 that $P = \zeta(3)$. That is, $Q = \int_0^1 \int_0^1 \int_0^1 \frac{dxdydz}{1+xyz} = \frac{3}{4}\zeta(3)$.

Challenge Problem 2.5.1: The two limiting operations are from different situations. When we say the harmonic series diverges logarithmically, we mean that the partial sums of a finite number of terms increase as the logarithm of the number of terms used. When we say $\zeta(s)$ diverges hyperbolically as $s \to 1$, we are speaking of the behavior of the total sum (of an infinite number of terms) as $s \to 1$.

Challenge Problem 2.6.1: From (2.1.9) we have $\eta(s) = \sum_{n=1}^{\infty} \frac{(-1)^{n+1}}{n^s}$ and from (1.4.21) we have $\int_0^{\infty} e^{-nx} x^{s-1} dx = \frac{\Gamma(s)}{n^s}$, and so $\frac{1}{n^s} = \frac{1}{\Gamma(s)}\int_0^{\infty} e^{-nx} x^{s-1} dx$. Thus, $\eta(s) = \sum_{n=1}^{\infty} \frac{(-1)^{n+1}}{n^s} = \frac{1}{\Gamma(s)}\sum_{n=1}^{\infty}(-1)^{n+1}\int_0^{\infty} e^{-nx} x^{s-1} dx$, or $\eta(s)\Gamma(s) = \sum_{n=1}^{\infty}(-1)^{n+1}\int_0^{\infty} e^{-nx} x^{s-1} dx = \int_0^{\infty} x^{s-1}\sum_{n=1}^{\infty}(-1)^{n+1} e^{-nx} dx = \int_0^{\infty} x^{s-1} T dx$, where $T = \sum_{n=1}^{\infty}(-1)^{n+1} e^{-nx} = e^{-x} - e^{-2x} + e^{-3x} - \cdots$. So $e^{-x}T = e^{-2x} - e^{-3x} + \cdots$, which says that $e^{-x}T + T = e^{-x} = T(e^{-x} + 1)$, or $T = \frac{e^{-x}}{e^{-x}+1} = \frac{1}{1+e^x}$ and therefore $\eta(s)\Gamma(s) = \int_0^{\infty} \frac{x^{s-1}}{e^x+1} dx$.

Challenge Problem 2.6.2: From (2.6.1) we have $\zeta(s) = \frac{1}{1-2^{1-s}}\sum_{k=1}^{\infty} \frac{(-1)^{k+1}}{k^s}$. Thus, $\zeta(\tfrac{1}{2}+ib) = \frac{1}{1-2^{\frac{1}{2}-ib}}\sum_{k=1}^{\infty} \frac{(-1)^{k+1}}{k^{\frac{1}{2}+ib}} = 0$. Clearly, $1 - 2^{\frac{1}{2}-ib} \neq \infty$ for any real value of b, and so it must be that $\sum_{k=1}^{\infty} \frac{(-1)^{k+1}}{k^{\frac{1}{2}+ib}} = 0$. Now, by Euler's identity $\frac{1}{k^{\frac{1}{2}+ib}} = \frac{k^{-ib}}{\sqrt{k}} = \frac{e^{\ln(k^{-ib})}}{\sqrt{k}} = \frac{1}{\sqrt{k}} e^{-ib\ln(k)} = \frac{1}{\sqrt{k}}[\cos\{b\ln(k)\} - i\sin\{b\ln(k)\}] = 0$. Thus, for each value of b such that $\zeta(\frac{1}{2}+ib) = 0$, it must be true that $\sum_{k=1}^{\infty} \frac{(-1)^{k+1}}{\sqrt{k}} \cos\{b\ln(k)\} = 0$ as well as $\sum_{k=1}^{\infty} \frac{(-1)^{k+1}}{\sqrt{k}} \sin\{b\ln(k)\} = 0$.

Challenge Problem 2.6.3: From the reflection formula for the gamma function, (1.4.20), we have $\Gamma(n)\Gamma(1-n) = \frac{\pi}{\sin(n\pi)}$. On the critical line $n\,(=s) = \frac{1}{2}+ib$, and so $\Gamma(n) = \Gamma(\frac{1}{2}+ib)$ and $\Gamma(1-n) = \Gamma(\frac{1}{2}-ib)$. Notice that this says that on the critical line, $\Gamma(n)$ and $\Gamma(1-n)$ are a complex conjugate pair. That is, $\Gamma(\frac{1}{2}+ib)\Gamma(\frac{1}{2}-ib) = |\Gamma(\frac{1}{2}+ib)|^2$ and this equals $\left|\frac{\pi}{\sin(\{\frac{1}{2}+ib)\pi\}}\right| = \frac{\pi}{|\sin(\{\frac{1}{2}+ib)\pi\}|}$. Now, from Euler's identity, $\sin\{(\frac{1}{2}+ib)\pi\} = \frac{e^{i((\frac{1}{2}+ib)\pi)} - e^{-i((\frac{1}{2}+ib)\pi)}}{2i}$ $= \frac{e^{i\pi/2}e^{-\pi b} - e^{-i\pi/2}e^{\pi b}}{2i} = \frac{ie^{-\pi b} + ie^{\pi b}}{2i} = \frac{e^{-\pi b} + e^{\pi b}}{2} = \cosh(\pi b)$. Since $\cosh(\pi b)$ is never negative for any real b, we can drop the absolute value sign and write $\Gamma(s)|_{s=\frac{1}{2}+ib} = \sqrt{\frac{\pi}{\cosh(\pi b)}}$, which is never zero for any real b.

Challenge Problem 2.7.1: (1) For pq to be perfect, we require $pq = 1 + p + q$. But since $(p-1)(q-1) > 2$ (because the smallest values that p and q can take on are 3 and 5), we have $pq - q - p + 1 > 2$ or $pq > q + p + 1$, which means pq is not perfect. (2) Following the hint, the divisors of p^k are $1, p, p^2, p^3, \ldots, p^{k-1}$. To be perfect, $p^k = 1 + p + p^2 + p^3 + \cdots + p^{k-1} = S$. Since

$$S = 1 + p + p^2 + p^3 + \cdots + p^{k-1}$$

then

$$pS = p + p^2 + p^3 + \cdots + p^{k-1} + p^k$$

and so $S - pS = 1 - p^k$, or $S = \frac{1-p^k}{1-p}$. Thus, we require $p^k = \frac{1-p^k}{1-p}$, or with just a bit of easy algebra, $2 - p = \frac{1}{p^k}$, which is impossible for any prime p (for $p \geq 2$, we have $2 - p \leq 0$ while $\frac{1}{p^k} > 0$). (3) A perfect square has the form s^2. An even perfect number has the form $2^{p-1}(2^p - 1)$, where $2^p - 1$ is prime (cannot be factored). Thus, to be a perfect square, it must be true that $s = 2^p - 1$ (thus $s = 2^{p-1}$, too) and so, with a bit of easy algebra, $2^p = 2$. But there is no integer $p > 1$ (which means no primes) where this is true.

Challenge Problem 3.1.1: By inspection of Figure 3.1.3, after n iterations there are 4^n line segments, each of length $(\frac{1}{3})^n$. The total length of the von Koch curve is therefore $(\frac{4}{3})^n$. The number of inches in a light-year is $12 \times 5{,}280 \times 186{,}000 \times 365 \times 24 \times 60 \times 60 = N = 3.7165 \times 10^{17}$. We require the first value of n such that $(\frac{4}{3})^n \geq N$ and so $n \geq \frac{\ln(N)}{\ln(\frac{4}{3})} = \frac{40.456}{0.28768} = 140.627$. Since n is an integer, $n = 141$ iterations. The number of line segments is $4^{141} = 7.77 \times 10^{84}$, a number that is greater than the total number of elementary particles in the entire observable universe.

Challenge Problem 3.2.1: (a) Each of $\cos(t)$ and $\cos(t\sqrt{2})$ is individually periodic: $\cos(t)$ has period 2π and $\cos(t\sqrt{2})$ has period $\frac{2\pi}{\sqrt{2}}$. Their sum, however, is not periodic. Here's why. Suppose there is a T such that $\cos(t) + \cos(t\sqrt{2}) = \cos(t+T) + \cos\{(t+T)\sqrt{2}\}$. For this to be true for all t, it must be true in particular for $t = 0$. This says $\cos(0) + \cos(0) = 2 = \cos(T) + \cos\{T\sqrt{2}\}$. Now, the maximum value of the cosine function is 1, and this says $\cos(T) = \cos\{T\sqrt{2}\} = 1$. That is, there must be two (obviously different) integers m and n such that $T = 2\pi n$ and $T\sqrt{2} = 2\pi m$, which says that $\frac{T\sqrt{2}}{T} = \frac{2\pi m}{2\pi n} = \sqrt{2} = \frac{m}{n}$, implying that $\sqrt{2}$ is rational. But it is well known that $\sqrt{2}$ is irrational. This contradiction means that our original assumption, that the sum is periodic, must be invalid. (b) For N any positive integer, define the two periodic functions $x_1(t)$ and $x_2(t)$ as follows: $x_1(t) = \begin{cases} \sin(2N\pi t), 0 \leq t \leq \frac{1}{N} \\ 0, \frac{1}{N} < t < 1 \end{cases}$ and $x_2(t) = \begin{cases} 0, 0 \leq t \leq \frac{1}{N} \\ \sin(2N\pi t), \frac{1}{N} < t < 1 \end{cases}$. That is, $x_1(t)$ and $x_2(t)$ have the same period of 1, independent of N. Their sum is $x_1(t) + x_2(t) = \sin(2N\pi t)$, $0 \leq t < 1$, which is periodic with period T, where $2N\pi T = 2\pi$. That is, $T = \frac{1}{N}$. Thus, while $x_1(t)$ and $x_2(t)$ each have period 1 for any N, by picking N arbitrarily large, we can make the period of their sum as small as we wish!

Challenge Problem 3.3.1: Following the hint, with $t = \pi$ and $\alpha = \frac{1}{2}$ in (3.3.14), we have $\cos(\frac{\pi}{2}) = \frac{\sin(\frac{\pi}{2})}{\pi}[2 + \sum_{k=1}^{\infty} \frac{(-1)^k}{\frac{1}{4} - k^2} \cos(k\pi)]$. Since

$\cos(\frac{\pi}{2}) = 0$, $\sin(\frac{\pi}{2}) = 1$, and $\cos(k\pi) = (-1)^k$, then all this reduces to $2 + \Sigma_{k=1}^{\infty} \frac{4}{1-4k^2} = 0$ or $\Sigma_{k=1}^{\infty} \frac{1}{1-4k^2} = -\frac{1}{2}$. Alternatively, for a second, purely algebraic way to sum the series, write $\Sigma_{k=1}^{\infty} \frac{1}{1-4k^2} = \Sigma_{k=1}^{\infty} \frac{1}{(1-2k)(1+2k)}$ $= \frac{1}{2}\Sigma_{k=1}^{\infty}\{\frac{1}{(1+2k)} + \frac{1}{(1-2k)}\} = \frac{1}{2}[\{-1 + \frac{1}{3}\} + \{-\frac{1}{3} + \frac{1}{5}\} + \{-\frac{1}{5} + \frac{1}{7}\} + \cdots]$ and then notice that all the terms but the first "telescope" to immediately give a sum of $\frac{1}{2}[-1] = -\frac{1}{2}$.

Challenge Problem 3.3.2: $(\frac{\pi}{2p})\frac{1+e^{-2\pi p}}{1-e^{-2\pi p}} - \frac{1}{2p^2} = (\frac{\pi}{2p})\frac{1 + [1 - 2\pi p + \frac{4\pi^2 p^2}{2} - \frac{8\pi^3 p^3}{6} + \cdots]}{1 - [1 - 2\pi p + \frac{4\pi^2 p^2}{2} - \frac{8\pi^3 p^3}{6} + \cdots]} - \frac{1}{2p^2}$.
As $p \to 0$ this becomes, with decreasing error,

$$(\frac{\pi}{2p})\frac{2 - 2\pi p + 2\pi^2 p^2 - \frac{4\pi^3 p^3}{3}}{2\pi p - 2\pi^2 p^2 + \frac{4\pi^3 p^3}{3}} - \frac{1}{2p^2} = (\frac{1}{p})\frac{1 - \pi p + \pi^2 p^2 - \frac{2\pi^3 p^3}{3}}{2p - 2\pi p^2 + \frac{4\pi^2 p^3}{3}} - \frac{1}{2p^2}$$

$$= (\frac{1}{2p^2})\frac{1 - \pi p + \pi^2 p^2 - \frac{2\pi^3 p^3}{3}}{1 - \pi p + \frac{2\pi^2 p^2}{3}} - \frac{1}{2p^2} = (\frac{1}{2p^2})\left[\frac{1 - \pi p + \pi^2 p^2 - \frac{2\pi^3 p^3}{3}}{1 - \pi p + \frac{2\pi^2 p^2}{3}} - 1\right]$$

$$= (\frac{1}{2p^2})\left[\frac{1 - \pi p + \pi^2 p^2 - \frac{2\pi^3 p^3}{3} - 1 + \pi p - \frac{2\pi^2 p^2}{3}}{1 - \pi p + \frac{2\pi^2 p^2}{3}}\right] = (\frac{1}{2p^2})\left[\frac{p^2(\pi^2 - \frac{2\pi^2}{3}) - \frac{2\pi^3 p^3}{3}}{1 - \pi p + \frac{2\pi^2 p^2}{3}}\right]$$

$$= \frac{\frac{\pi^2}{3} - \frac{2\pi^3 p}{3}}{2(1 - \pi p + \frac{2\pi^2 p^2}{3})} \to \frac{\pi^2}{6}$$

as $p \to 0$.

Challenge Problem 3.3.3: The Fourier coefficients are $a_m = \frac{2}{T}\int_{-T/2}^{T/2} f(t)\cos(mt)dt$ and $b_m = \frac{2}{T}\int_{-T/2}^{T/2} f(t)\sin(mt)dt$. The convergence argument in note 9 of Chapter 3 says that $\lim_{m\to\infty} a_m = \lim_{m\to\infty} b_m = 0$, which requires that the Fourier integrals vanish as $m \to \infty$, and that is the Riemann-Lebesgue lemma.

Challenge Problem 3.4.1: From (2.1.9) and (2.1.10) we have $\eta(s) = \Sigma_{k=1}^{\infty}\frac{(-1)^{k+1}}{k^s} = \frac{1}{1^s} - \frac{1}{2^s} + \frac{1}{3^s} - \frac{1}{4^s} + \cdots = (1 - 2^{1-s})\zeta(s)$, or with $s = 4$, $1 - \frac{1}{2^4} + \frac{1}{3^4} - \frac{1}{4^4} + \cdots = (1 - 2^{-3})\zeta(4) = (\frac{7}{8})\frac{\pi^4}{90} = \frac{7\pi^4}{720}$.

Challenge Problem 3.4.2: Setting $t = 5$ in (3.4.17), we have $\Sigma_{k=1}^{4}\zeta(2k)\zeta(10 - 2k) = \frac{11}{2}\zeta(10) = \zeta(2)\zeta(8) + \zeta(4)\zeta(6) + \zeta(6)\zeta(4) + \zeta(8)\zeta(2)$ $= 2[\zeta(2)\zeta(8) + \zeta(4)\zeta(6)]$

$$= 2[(\tfrac{\pi^6}{6})(\tfrac{\pi^8}{9,450}) + (\tfrac{\pi^4}{90})(\tfrac{\pi^6}{945})] = 2\pi^{10}[(\tfrac{1}{6})(\tfrac{1}{945})\{\tfrac{1}{10} + \tfrac{1}{15}\}]$$

$$= 2\pi^{10}[(\tfrac{1}{6})(\tfrac{1}{945})(\tfrac{1}{5})\{\tfrac{1}{2} + \tfrac{1}{3}\}]$$

$$= 2\pi^{10}[(\tfrac{1}{6})(\tfrac{1}{945})(\tfrac{1}{5})(\tfrac{5}{6})] = \pi^{10}[(\tfrac{1}{6})(\tfrac{1}{945})(\tfrac{1}{3})] = \tfrac{11}{2}\zeta(10).\ \text{So}$$

$$\zeta(10) = \pi^{10}\tfrac{2}{11}[(\tfrac{1}{6})(\tfrac{1}{945})(\tfrac{1}{3})] = \tfrac{\pi^{10}}{(99)(945)} = \tfrac{\pi^{10}}{93,555}.$$

Challenge Problem 3.5.1: Following the hint, putting $t = \tfrac{\pi}{4}$ into (2.3.1) gives

$$\tfrac{\pi - \tfrac{\pi}{4}}{2} = \tfrac{3\pi}{8} = \sin(\tfrac{\pi}{4}) + \tfrac{\sin(\tfrac{\pi}{4})}{2} + \tfrac{\sin(\tfrac{3\pi}{4})}{3} + \tfrac{\sin(\pi)}{4} + \tfrac{\sin(\tfrac{5\pi}{4})}{5} + \tfrac{\sin(\tfrac{6\pi}{4})}{6} + \tfrac{\sin(\tfrac{7\pi}{4})}{7} + \tfrac{\sin(2\pi)}{8} + \tfrac{\sin(\tfrac{9\pi}{4})}{9} + \cdots,$$

or $\tfrac{3\pi}{8} = \tfrac{1}{\sqrt{2}} + \tfrac{1}{2} + \tfrac{1}{3\sqrt{2}} - \tfrac{1}{5\sqrt{2}} - \tfrac{1}{6} - \tfrac{1}{7\sqrt{2}} + \tfrac{1}{9\sqrt{2}} + \tfrac{1}{10} + \tfrac{1}{11\sqrt{2}} + \cdots,$ or

$$\tfrac{3\pi}{8} = \tfrac{1}{\sqrt{2}} + \tfrac{1}{\sqrt{2}\sqrt{2}} + \tfrac{1}{3\sqrt{2}} - \tfrac{1}{5\sqrt{2}} - \tfrac{1}{3\sqrt{2}\sqrt{2}} - \tfrac{1}{7\sqrt{2}} + \tfrac{1}{9\sqrt{2}} + \tfrac{1}{5\sqrt{2}\sqrt{2}} + \tfrac{1}{11\sqrt{2}} + \cdots, \text{ or}$$

$$\tfrac{3\pi\sqrt{2}}{8} = 1 + \tfrac{1}{\sqrt{2}} + \tfrac{1}{3} - \tfrac{1}{5} - \tfrac{1}{3\sqrt{2}} - \tfrac{1}{7} + \tfrac{1}{9} + \tfrac{1}{5\sqrt{2}} + \tfrac{1}{11} + \cdots, \text{ or}$$

$$\tfrac{3\pi\sqrt{2}}{8} = 1 + \tfrac{1}{3} - \tfrac{1}{5} - \tfrac{1}{7} + \tfrac{1}{9} + \tfrac{1}{11} - \cdots + [\tfrac{1}{\sqrt{2}} - \tfrac{1}{3\sqrt{2}} + \tfrac{1}{5\sqrt{2}} - \cdots], \text{ or}$$

$$\tfrac{3\pi\sqrt{2}}{8} = 1 + \tfrac{1}{3} - \tfrac{1}{5} - \tfrac{1}{7} + \tfrac{1}{9} + \tfrac{1}{11} - \cdots + \tfrac{1}{\sqrt{2}}[1 - \tfrac{1}{3} + \tfrac{1}{5} - \cdots], \text{ or recalling the}$$ Gregory/Leibniz series, $\tfrac{3\pi\sqrt{2}}{8} = 1 + \tfrac{1}{3} - \tfrac{1}{5} - \tfrac{1}{7} + \tfrac{1}{9} + \tfrac{1}{11} - \cdots + \tfrac{1}{\sqrt{2}}(\tfrac{\pi}{4}).$ Now, $\tfrac{3\pi\sqrt{2}}{8} - \tfrac{\pi}{4\sqrt{2}} = \tfrac{3\pi\sqrt{2}}{8} - \tfrac{\pi\sqrt{2}}{8} = \tfrac{2\pi\sqrt{2}}{8}$ and so $\tfrac{\pi\sqrt{2}}{4} = 1 + \tfrac{1}{3} - \tfrac{1}{5} - \tfrac{1}{7} + \tfrac{1}{9} + \tfrac{1}{11} - - \cdots.$

Challenge Problem 3.6.1: Starting with $1 + 2\sum_{k=1}^{\infty} e^{-k^2} = \sqrt{\pi}(1 + 2\sum_{n=1}^{\infty} e^{-\pi^2 n^2})$, then coding just the first three terms of the left-hand side gives $1.772636\ldots$, while coding just the first three terms of the right-hand side gives $1.772637\ldots$. As claimed, we don't see disagreement until the sixth decimal place.

Challenge Problem 3.6.2: The total energy of $f(t)$ is $\int_{-\infty}^{\infty} f^2(t)dt = \int_0^1 dt = 1$. So, if $F(\omega)$ is the Fourier transform of $f(t)$, then $\int_0^{\infty} \tfrac{1}{\pi}|F(\omega)|^2\, d\omega = 1$. Now, $F(\omega) = \int_{-\infty}^{\infty} f(t)e^{-i\omega t} dt = \int_0^1 e^{-i\omega t} dt = \tfrac{e^{-i\omega t}}{-i\omega}\big|_0^1 = \tfrac{e^{-i\omega}-1}{-i\omega} = \tfrac{e^{-\tfrac{i\omega}{2}}(e^{-\tfrac{i\omega}{2}} - e^{\tfrac{i\omega}{2}})}{-i2\tfrac{\omega}{2}} = \tfrac{-2i\sin(\tfrac{\omega}{2})e^{-\tfrac{i\omega}{2}}}{-2i\tfrac{\omega}{2}}$ $= \tfrac{\sin(\tfrac{\omega}{2})e^{-\tfrac{i\omega}{2}}}{\tfrac{\omega}{2}}.$ So $F^*(\omega) = \tfrac{\sin(\tfrac{\omega}{2})e^{\tfrac{i\omega}{2}}}{\tfrac{\omega}{2}}$ and $F(\omega)F^*(\omega) = |F(\omega)|^2 = \tfrac{\sin^2(\tfrac{\omega}{2})}{(\tfrac{\omega}{2})^2}$ and so

$\int_0^\infty \frac{1}{\pi} \frac{\sin^2(\frac{\omega}{2})}{(\frac{\omega}{2})^2} d\omega = 1$, or $\int_0^\infty \frac{\sin^2(\frac{\omega}{2})}{(\frac{\omega}{2})^2} d\omega = \pi$. Now let $\frac{\omega}{2} = ax$ (and so $d\omega = 2adx$).

Then $\int_0^\infty \frac{\sin^2(ax)}{a^2 x^2} 2a\,dx = \pi = 2\int_0^\infty \frac{\sin^2(ax)}{ax^2} dx$ or $\int_0^\infty \frac{\sin^2(ax)}{x^2} dx = \frac{\pi a}{2}$.

Challenge Problem 3.6.3: (a) We need only consider what $\varphi(t)$ is doing at $t = 0$, because for $t \neq 0$, $\delta(t)\varphi(t) = 0$. Suppose, then, we look at $\varphi(t)$ in the interval $-\varepsilon < t < \varepsilon$, where $\varepsilon > 0$, and then imagine $\varepsilon \to 0$. Over that interval $\varphi(t)$ starts at $\varphi(-\varepsilon)$ and ends at $\varphi(\varepsilon)$, but as t varies, we argue that $\varphi(t)$ can't vary by much, since $\varphi(t)$ is given as "smoothly varying." That is, $\varphi(t)$ must remain pretty close to $\varphi(0)$ over the entire integration interval as $\varepsilon \to 0$. So $\int_{-\infty}^\infty \delta(t)\varphi(t)dt$ $= \lim_{\varepsilon \to 0} \int_{-\varepsilon}^\varepsilon \delta(t)\varphi(t)dt = \lim_{\varepsilon \to 0} \int_{-\varepsilon}^\varepsilon \delta(t)\varphi(0)dt = \varphi(0)\lim_{\varepsilon \to 0} \int_{-\varepsilon}^\varepsilon \delta(t)dt = \varphi(0)$ as the unit area of $\delta(t)$ is all located at $t = 0$. This is fast-and-loose math, yes, and I admit physicists and engineers are much happier with this sort of argument than are mathematicians. (b) $F(\omega) = \int_{-\infty}^\infty \delta(t)e^{-i\omega t}dt = e^{-i\omega t}|_{t=0} = 1$. (c) $\frac{1}{\pi}|F(\omega)|^2 = \frac{1}{\pi}$ and so $W = \int_0^\infty \frac{1}{\pi} d\omega = \infty$. That is, the Dirac impulse function is an infinite energy function, even though it is identically zero except for one instant of time. The infinite energy of the Dirac impulse is delivered in a single instant, so it's one mighty blast—don't get in its way!

Challenge Problem 3.7.1: $\zeta(s) = \Sigma_{k=1}^\infty \frac{1}{k^s} = \zeta(x+it) = \Sigma_{k=1}^\infty \frac{1}{k^{x+it}}$ $= \Sigma_{k=1}^\infty \frac{1}{k^x k^{it}} = \Sigma_{k=1}^\infty \frac{1}{k^x e^{\ln(k^{it})}} = \Sigma_{k=1}^\infty \frac{e^{-it\ln(k)}}{k^x}$. Or using Euler's identity, $\zeta(x+it)$ $= \Sigma_{k=1}^\infty \frac{\cos\{t\ln(k)\}}{k^x} - i\Sigma_{k=1}^\infty \frac{\sin\{t\ln(k)\}}{k^x}$. These two sums converge if $x > 1$, and each is easily coded. Using the first 1 million terms of each sum (with $x = 2$ and $t = 1$), we get $\zeta(2 + i) = 1.150356\ldots - i\,0.43753\ldots$. As a check, the *MATLAB* function *zeta* calculates *zeta*$(2 + i) = 1.150355 \ldots - i\,0.43753\ldots$, in excellent agreement.

Challenge Problem 3.7.2: Since $\zeta(s) = 2(2\pi)^{s-1}\sin(\frac{\pi s}{2})\Gamma(1-s)\zeta(1-s)$, then with $s = -\frac{1}{2}$ we have $\zeta(-\frac{1}{2}) = 2(2\pi)^{-\frac{3}{2}}\sin(-\frac{\pi}{4})\Gamma(\frac{3}{2})\zeta(\frac{3}{2})$ $= \frac{2}{2\pi\sqrt{2\pi}}(-\frac{1}{\sqrt{2}})\Gamma(\frac{3}{2})\zeta(\frac{3}{2}) = -\frac{\Gamma(\frac{3}{2})\zeta(\frac{3}{2})}{2\pi\sqrt{\pi}}$. From (1.4.3) and (1.4.6), $\Gamma(n + 1) = n\Gamma(n)$, or with $n = \frac{1}{2}$, $\Gamma(\frac{3}{2}) = \frac{1}{2}\Gamma(\frac{1}{2}) = \frac{1}{2}\sqrt{\pi}$ and so $\zeta(-\frac{1}{2}) = -\frac{\frac{1}{2}\sqrt{\pi}}{2\pi\sqrt{\pi}}\zeta(\frac{3}{2})$ $= -\frac{1}{4\pi}\zeta(\frac{3}{2}) = -\frac{1}{4\pi}\Sigma_{k=1}^\infty \frac{1}{k^{3/2}}$. This is easy to code, and using the first 1 million terms of the series gives $\zeta(-\frac{1}{2}) = -0.207727\ldots$, which is consistent with the lower-right plot of Figure 3.7.1. Indeed, *MATLAB* calculates $zeta(-\frac{1}{2}) = -0.207886\ldots$.

Challenge Problem 4.1.1: $\frac{5}{2}\zeta(4) - \frac{1}{2}\zeta^2(2) = \frac{5}{2}(\frac{\pi^4}{90}) - \frac{1}{2}(\frac{\pi^2}{6})^2 = \frac{5\pi^4}{180} - \frac{\pi^4}{72} = \frac{\pi^4}{36}$ $-\frac{\pi^4}{72} = \frac{2\pi^4}{72} - \frac{\pi^4}{72} = \frac{\pi^4}{72}$, as given in (4.1.2).

Challenge Problem 4.2.1: Following the hint, we start with Euler's (4.1.2) and write $\frac{5}{4}\zeta(4) = \sum_{q=1}^{\infty} \frac{h(q)}{q^3} = \sum_{q=1}^{\infty} \frac{h(q-1)+\frac{1}{q}}{q^3} = \sum_{q=1}^{\infty} \frac{h(q-1)}{q^3} + \sum_{q=1}^{\infty} \frac{1}{q^4}$. Since the last sum is $\zeta(4)$, we have $\frac{1}{4}\zeta(4) = \sum_{q=1}^{\infty} \frac{h(q-1)}{q^3}$. Now change the index to $q-1 = k$ (and so $q = k+1$) to get $\frac{1}{4}\zeta(4) = \sum_{k=0}^{\infty} \frac{h(k)}{(k+1)^3} = h(0) + \sum_{k=1}^{\infty} \frac{h(k)}{(k+1)^3}$. Since $h(0) = 0$, we instantly have our answer: $\sum_{q=1}^{\infty} \frac{h(q)}{(q+1)^3} = \frac{1}{4}\zeta(4)$. Since $\zeta(4) = \frac{\pi^4}{90}$, we have $\frac{1}{4}\zeta(4) = \frac{\pi^4}{360} = 0.27058\ldots$, which agrees pretty well with the numerical estimate produced by **euler1775.m**.

Challenge Problem 4.2.2: Following the hint, in $\int_0^x \frac{dt}{1-t}$ let $u = 1 - t$ (and so $dt = -du$). Then $\int_0^x \frac{dt}{1-t} = \int_1^{1-x} \frac{-du}{u} = \int_{1-x}^1 \frac{du}{u} = \ln(u)\big|_{1-x}^1 = -\ln(1-x)$, that is, $\ln(1-x) = -\int_0^x \frac{dt}{1-t}$. So $\int_0^1 x^{q-1}\ln(1-x)dx = \int_0^1 x^{q-1}\{-\int_0^x \frac{dt}{1-t}\}dx = -\int_0^1 \frac{1}{1-t}\{\int_t^1 x^{q-1}dx\}dt$, where I've used the remarks made in the Special Note of the problem statement. Now, since $\int_t^1 x^{q-1}dx = \frac{x^q}{q}\big|_t^1 = \frac{1}{q} - \frac{t^q}{q} = \frac{1-t^q}{q}$, then $\int_0^1 x^{q-1}\ln(1-x)dx$ $-\frac{1}{q}\int_0^1 \frac{1-t^q}{1-t}dt$. Since $\frac{1-t^q}{1-t} = (1-t^q)(1 + t + t^2 + t^3 + \cdots) = (1 + t + t^2 + t^3 + \cdots)$ $-(t^q + t^{q+1} + t^{q+2} + \cdots) = 1 + t + t^2 + t^3 + \cdots + t^{q-1}$, we have $\int_0^1 x^{q-1}\ln(1-x)dx$ $= -\frac{1}{q}\int_0^1 (1 + t + t^2 + t^3 + \cdots + t^{q-1})dt = -\frac{1}{q}(t + \frac{t^2}{2} + \frac{t^3}{3} + \cdots + \frac{t^q}{q})\big|_0^1 = -\frac{1}{q}(1 + \frac{1}{2} + \frac{1}{3} + \cdots + \frac{1}{q})$ $= -\frac{h(q)}{q}$.

Challenge Problem 4.3.1: The identity to be shown is equivalent to saying $\sum_{q=1}^{\infty} x^q(1-x)h(q) = -\ln(1-x)$. So expanding the left-hand side, $\sum_{q=1}^{\infty} x^q(1-x)h(q) = \sum_{q=1}^{\infty} x^q h(q) - \sum_{q=1}^{\infty} x^{q+1}h(q) = \sum_{q=1}^{\infty} x^q h(q) - \sum_{q=1}^{\infty} x^{q+1}\{h(q+1)$ $-\frac{1}{q+1}\} = \sum_{q=1}^{\infty} x^q h(q) - \sum_{q=1}^{\infty} x^{q+1}h(q+1) + \sum_{q=1}^{\infty} \frac{x^{q+1}}{q+1} = \lim_{N \to \infty}\{\sum_{q=1}^N x^q h(q) - \sum_{q=1}^N x^{q+1}$ $h(q+1)\} + \sum_{q=1}^{\infty} \frac{x^{q+1}}{q+1}$. Now, $\sum_{q=1}^N x^q h(q) = xh(1) + x^2 h(2) + x^3 h(3) + \cdots + x^N h(N)$, and $\sum_{q=1}^N x^{q+1}h(q+1) = x^2 h(2) + x^3 h(3) + \cdots + x^N h(N) + x^{N+1}h(N+1)$. Thus, $\sum_{q=1}^N x^q h(q) - \sum_{q=1}^N x^{q+1}h(q+1) = xh(1) - x^{N+1}h(N+1)$, or as $h(1) = 1$, $\sum_{q=1}^N x^q h(q)$ $-\sum_{q=1}^N x^{q+1}h(q+1) = x - x^{N+1}h(N+1)$. Also, $\sum_{q=1}^{\infty} \frac{x^{q+1}}{q+1} = \frac{x^2}{2} + \frac{x^3}{3} + \frac{x^4}{4} + \cdots$. From (1.3.5), $x = \ln(1+x) + \frac{x^2}{2} - \frac{x^3}{3} + \frac{x^4}{4} - \cdots$, or replacing x with $-x$, $-x$ $= \ln(1-x) + \frac{x^2}{2} + \frac{x^3}{3} + \frac{x^4}{4} + \cdots$, and so $\frac{x^2}{2} + \frac{x^3}{3} + \frac{x^4}{4} + \cdots = -x - \ln(1-x)$ and we have $\sum_{q=1}^{\infty} \frac{x^{q+1}}{q+1} = -x - \ln(1-x)$. Thus $\sum_{q=1}^{\infty} x^q(1-x)h(q) = \lim_{N \to \infty}\{x - x^{N+1}h(N+1)\}$

$-x - \ln(1-x) = -\lim_{N \to \infty} x^{N+1} h(N+1) - \ln(1-x)$. Now for $-1 < x < 1$, $\lim_{N \to \infty} x^{N+1} h(N+1) = 0$, and so as claimed, $\Sigma_{q=1}^{\infty} x^q (1-x) h(q) = -\ln(1-x)$, or $\Sigma_{q=1}^{\infty} x^q h(q) = -\frac{\ln(1-x)}{1-x}$, $-1 < x < 1$. Finally, as both sides of this identity obviously blow up at $x = 1$, we can formally extend $-1 < x < 1$ to $-1 < x \le 1$.

Final Challenge Problem: In De Doelder's identity $\Sigma_{q=1}^{\infty} \frac{h^2(q)}{(q+1)^2} = \frac{11}{4}\zeta(4)$, change the index in the sum to $k = q + 1$ ($q = k - 1$). Then $\Sigma_{k=2}^{\infty} \frac{h^2(k-1)}{k^2}$ $= \{\Sigma_{k=1}^{\infty} \frac{h^2(k-1)}{k^2} - \frac{h^2(0)}{1}\}$, or since $h(0) = 0$, we have $\Sigma_{q=1}^{\infty} \frac{h^2(q)}{(q+1)^2} = \Sigma_{q=1}^{\infty} \frac{h^2(q-1)}{q^2} = \frac{11}{4}\zeta(4)$. Now, $h^2(q-1) = [h(q) - \frac{1}{q}]^2 = h^2(q) - \frac{2}{q}h(q) + \frac{1}{q^2}$. Thus, $\Sigma_{q=1}^{\infty} \frac{h^2(q)}{(q+1)^2} = \Sigma_{q=1}^{\infty} \{\frac{h^2(q)}{q^2} -\frac{2}{q^3}h(q) + \frac{1}{q^4}\} = \frac{11}{4}\zeta(4)$, and so $\Sigma_{q=1}^{\infty} \frac{h^2(q)}{q^2} = \frac{11}{4}\zeta(4) + 2\Sigma_{q=1}^{\infty} \frac{h(q)}{q^3} - \Sigma_{q=1}^{\infty} \frac{1}{q^4}$, or recalling Euler's (4.1.2), we immediately have $\Sigma_{q=1}^{\infty} \frac{h^2(q)}{q^2} = \frac{11}{4}\zeta(4) + 2\frac{5}{4}\zeta(4)$ $-\zeta(4) = \frac{11}{4}\zeta(4) + \frac{10}{4}\zeta(4) - \zeta(4) = \frac{17}{4}\zeta(4)$, which was Au-Yeung's conjecture.

Acknowledgments

This book was difficult for me to write, requiring long periods of time reading tersely written technical papers before I reached a level of understanding that enabled me to explain (I hope!) it all to readers. But still, even with all that effort, none of it would have mattered without the support of various highly skilled and talented people. In particular, Susannah Shoemaker (my editor at Princeton University Press) and her efficient assistant Kristen Hop, Debbie Tegarden (the book's production editor, with whom I've spent more time talking on the telephone than with my own wife), and Cyd Westmoreland (who did a marvelous job of copyediting). Closer to home, I owe a debt of thanks to my grandchildren Erik and Lauren Stephens, who introduced me (during a riotous dinner conversation a couple of years ago) to the fun of Chuck Norris math jokes. As with the last two of my books, this one was written in the cozy Me & Ollie's Bakery & Café on Water Street in Exeter, New Hampshire. There is nothing that inspires mathematical writing like a comfortable, overstuffed leather chair, combined with a hot cup of coffee (and maybe a donut, too), while on the other side of the window a snowstorm roars. There isn't a faculty lounge in America that could have done a better job of providing the required scholarly ambience. Most of all, however, I owe everything to my wife of sixty years, Patricia Ann, who has stayed with me despite my habit of, now and then, insisting on singing the classic 1956 song *Jim Dandy to the Rescue* to her in bed at night.

Index

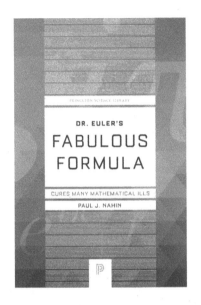
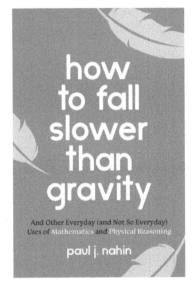